THE INNOVATION
ETHIC

THE INNOVATION ETHIC

Robert Kirk Mueller

American Management Association, Inc.

Foreword

Businessmen live in an environment — actually in a vortex — of change. How to deal with change in a rational way is a challenging, daily problem.

An important contributor to the rate of change today's businessman faces is the fact that he lives in a global village — any part of the world is within immediate reach by communication satellite or airplane. Businessmen now travel from continent to continent just as they traveled from city to city a generation ago. This ready access to all parts of the world has also contributed to the development of multinational corporations, which some view to be the dinosaurs of our time. Can the small brain of such a sprawling corporation (that is, its corporate headquarters) effectively control the tentacles it has extended around the world? This is in doubt. Perhaps, therefore, the future belongs to the small, efficient, localized corporation, rather than to the multinational giants. This is one of the many problems with which the modern businessman must deal.

Taking a different stance, a writer to *Le Monde,* the distinguished French newspaper, recently asked why it is that mankind has been able to produce 25 pounds of TNT for every human being but not 25 pounds of grain. A good question. It deserves an answer. To an increasing extent governments and business recognize such questions, and more and more an awareness of the social responsibility of business itself is recognized. One aspect of this present concern is the growing recognition of the need to think not only in terms of the Gross National Product but also in terms of the Net National Product, for no nation's resources, no matter how abundant, are infinite in quantity.

And still another problem facing us all is that one-third of the world's population is consuming two-thirds of the earth's resources; this cannot continue without serious international consequences.

These are some of the problems that businessmen are grappling with more and more, and these are the problems that emerge, from time to time, in Robert Kirk Mueller's *The Innovation Ethic.* Certainly the abundance of literature being published should be helpful to managers attempting to cope with these problems. Business magazines, books, and professional literature inundate the business community daily—far more than can be read with care. Selectivity has become increasingly important.

Businessmen must seek out terse, authoritative statements on the most important problems and opportunities they face. This is why I believe this book is significant and timely. It is succinct and it does focus on a fundamental aspect of management—innovation. Bob deals extensively with innovation as a concept, as a national phenomenon, and as a concern of specific industries. In his mind, and in the minds of most thoughtful businessmen, innovation is the key to business survival and successful entrepreneurial enterprise. The business community must be aware of the potential implications of innovation and, in particular, of technological growth. It must know how to use technology efficiently, and in the interest of improving the lot of humanity. To begin with, it must understand innovation, the need for it, its potential usefulness, and how to manage it. Recognizing these important facts, Bob offers thoughtful counsel and guidance on such matters as how to organize for innovation, the chief executive's role in an innovative enterprise, and the relationship between innovation and venture capital.

Bob Mueller is an exceptional businessman—exceptional because he reads extensively and because he is a philosopher as well as a businessman. He has been a senior executive of one of the nation's largest corporations and a member of a number of boards of directors; he is now a consultant with Arthur D. Little, Inc. Hence he is sensitive to change, for he is well informed about and aware of the basic undercurrents in our daily business lives. Bob writes about innovation in an erudite, thoughtful, and compelling way. Businessmen — and statesmen — would do well to heed his words.

James M. Gavin
Chairman
Arthur D. Little, Inc.

Preface

THE manuscript for this book was pressure-tested during May 1970 in Austria at the Salzburg Seminar in American Studies, attended by 55 Fellows from 20 countries in Eastern and Western Europe. Not surprisingly, the subject of innovation is as topical in Europe as it is in management circles in the United States. The European executives, however, were more interested in the philosophical, moral, and ethical aspects of our discussions than their counterparts are in the United States. Hence the title, *The Innovation Ethic,* seems appropriate.

The development of the manuscript was somewhat like a recent *Saturday Review* cartoon: The disheveled author, sitting before his publisher, says, "It started out as a suicide note. Then I corrected the spelling of a few words; rearranged a sentence or two; became interested in the style; developed plot lines and added suspense; inserted a few flashbacks to my miserable childhood and, of course, many bedroom episodes from my formative and adult years. Never once did I dream I would wind up talking to you about subsidiary rights."* This book was generated during thirty-plus years of managing experience, including management of a number of technical development projects and new businesses in various locations. Never once during this time, however, did I dream of writing a book about them.

A keystone study concerning technological innovation in industry is a 1963 report by Arthur D. Little, Inc., to the Na-

* Reprinted by permission of *Saturday Review* and the cartoonist, John A. Ruge.

tional Science Foundation.[1] ADL's experience with hundreds of specific cases involving different aspects of technical innovation, and specifically with more than 100 technical audits over the previous 15 years, was the background for the study. The report covers five industrial areas: textiles, machine tools, construction, appliances, and semiconductors. Organizational patterns, internal problems, and industry characteristics were examined. In addition, the study included certain institutions such as the U.S. Department of Agriculture, wartime research and development organizations—for example, the Office of Scientific Research and Development (OSRD)—and industrial associations, as well as the efforts of certain foreign governments to stimulate and support technical innovations. ADL's continuing work with clients in many fields testifies to the significance of and interest in innovation as a management process and an evolving ethic.

Europe so far is lagging considerably in developing an innovation ethic of its own. However, the Continent is receiving a major boost from an educational standpoint if the Organisation for Economic Co-Operation and Development (OECD) Council's proposed International Institute for the Management of Technology proceeds to implement its plans, which are well under way. The new organization, established by intergovernmental convention as a nonstock, nonprofit institution, has its headquarters in a sixteenth-century building (a former convent) donated by the city of Milan. The Institute's principal objective is to provide advanced training of managers and teachers, together with facilities for associated research in the management of technological innovation. The operating budget is about $1.5 million for the first year, which will rise to about $3 million when the endeavor is in full swing—by 1974.

It is surprising that, with more entrepreneurial spirit characterizing the European rather than the American managerial style, innovative ventures have not advanced as fast in Europe as in the United States. This is owing in large

[1] Arthur D. Little, Inc., *Patterns and Problems of Technical Innovation in American Industry*, report to National Science Foundation, September 1963, PB 181573.

part to public and governmental attitudes and to less recognition by the European financial community of the special
needs and sources of venture capital. As Chapter 9 on venture capital describes, the aggressive venture capital movement in the United States has spawned rapid advances in
technology and is a major difference between U.S. innovation
trends and those in Europe and the rest of the world.

Although the process of innovation is not confined to
technology, the current wave of interest in it throughout the
world stems from the reaction to the technological breakthroughs of recent years. In certain fields, other nations have
equaled or surpassed the United States. Japan's electronics
technology, for example, is moving forward so rapidly that it
may soon outstrip that of the United States. The advanced
housing technology of certain European countries would undoubtedly assist the United States in solving some of its own
housing problems.

Innovation, of course, is not a modern phenomenon, but
is something that some managers are now explicitly examining as a process and as a concept. A professional viewpoint
is emerging—an innovation ethic. The discipline dealing
with what is a worthy innovation and what is not has stirred
management to take innovative and responsive action in
both technical and social ways to cope with some of the impact of technological innovation.

The purpose of this book, then, is to examine the intellectual component of innovation in the current mosaic of the
management process. While some managers may be able to
deal with innovation at a high level of abstraction, the real
problem is what the computer world calls "conversion," or
"transduction," or how to convert a chief executive officer's
interest in innovation into innovative action. There are no
arbitrary conventions yet for dealing with this evolving professional ethic, the process of innovation. Therefore an understanding of the concepts and discipline of the process is
essential as a component of the overall management function.

Paul N. Herzog was the indirect provocateur of this manuscript but should bear no responsibility for its content. It was

through his request and guidance that my thoughts on the subject crystallized into discussions at the Salzburg Seminar in American Studies, of which Mr. Herzog is president. Let me also pay tribute to my associate faculty members and the Fellows of Seminar Session 128, who, during our stay in Austria, filtered and clarified the points of view that have emerged in this book.

During a long industrial management career, a more recent second career in management consulting, and part-time executive trusteeships in several financial and educational organizations, I have been given ideas, viewpoints, and specific information for the book by many colleagues and friends. To name all who contributed in great and small ways would make a book in itself. I should, however, mention my colleagues at Arthur D. Little and the following friends in the United States and as indicated around the world who shaped my thinking by discussion and correspondence throughout the years.

Dr. S. T. Adarkar, Bombay; Dr. J. W. Barrett, London; Edward S. Boyle and Abraham Friedman, Tel Aviv; Dr. John R. Durland, Tokyo; Warren C. Feist; J. C. Garrells, 3rd, Dr. R. S. Gordon, R. F. Hansen, S. T. Harris; Hans Hartung, Zurich; Elliott Haynes, Dr. J. M. Juran, Donald H. Korn; Dr. T. J. Kinsella, Taiwan; Enrique Krag, Buenos Aires; Lucien Levy and Peter D. Wood, Johannesburg; John H. Lippincott, J. A. Morton, and Daniel Meinertzhagen, London; Alejandro Medina-Mora, Mexico City; J. E. Murphy, Sidney Musher, H. K. Nason, D. S. Plumb, A. G. Quackenbush; S. Radhakrishnan, New Delhi; Dr. C. E. Reed, Richard T. Schotte, John C. Sevey, A. R. Taylor, M. C. Throdahl, Donald H. Wheeler; Enrique Uhthoff, Mexico City; C. J. Wickham, London; Dr. Robert G. Wilson, Saudi Arabia; Dr. Everett M. Woodman.

Listing their names in no way implies that they endorse the views expressed in this book; it simply expresses my gratitude to them for sharing their thoughts.

Robert Kirk Mueller

Contents

PART ONE

Innovation: Its Importance for Ongoing Institutions

1
Innovation Concepts

Change is what you should prepare for in pros-
perity and hope for in adversity. *Anonymous*

INNOVATION is a vast subject; it means many things, both
tangible and intangible, to many people. A new idea, theory,
machine, tool, social arrangement, and behavioral pattern
all fall under the general subject of innovation. Dr. Howard
O. McMahon, president of Arthur D. Little, Inc., has pointed
out that innovation is a uniquely human quality and that
change, renewal, and rejuvenation are normal, healthy human tendencies.[1] But while innovation is a natural human
activity, it is planned and deliberate. It is also a continuous
process, for whenever innovation occurs, change results, and
those affected by the change must in turn innovate in order
to respond. Thus innovation leads to change, which leads to
the necessity for further innovation, which again leads to
change — a continuing cycle. An organization which does not
confront change, or believes that it need not innovate, stag-
nates, decays, and dies.

All innovations are changes, but not all changes are innovations. An innovation is a deliberate, novel, specific change

NOTE: References are cited in full at the end of the book.

3

aimed at accomplishing the goals of the system more effectively. Initially, it is not a complete transformation of the
system. At a later stage, innovation reaches its limits only at
a point where the identity of the establishment itself is
threatened. The temper of innovation is positive rather than
revolutionary or rebellious, for if it antagonizes and is unacceptable, by definition an innovation lapses. If it does satisfy
a want or nullify an annoyance, as the sociologists say, the
innovation gathers a following and is accepted. The seminal
aspect of innovation, therefore, is useful change. Innovation,
too, is relative — if it is new to the people involved, it is an
innovation.

Invention versus Innovation

Invention is concerned with creativity and discovery and
generally implies fabrication, mental or otherwise. It is not
necessarily useful, and as long as it is a discovery — that is, if
something new is brought into being — it is an invention in
one nonlegal sense of the word. In order to be legally
patentable, however, an invention "must represent true innovation and add to the sum of useful knowledge," according
to the U.S. Supreme Court (December 1969). It is not enough
that a product "filled a long felt want and has enjoyed commercial success," the Court declared. In other words, com-
bining old elements is not innovative. William E. Schuyler,
Jr., U.S. Commissioner of Patents, has said, "All of us
recognize the need to protect innovation — the invention of
lesser importance — in order to provide the incentive necessary to stimulate creative effort in that area and marketing of
the innovation." [2] These basic concepts of invention were
laid down more than 600 years ago and were once characterized by Abraham Lincoln as having "added the fuel of
interest to the fire of genius."

The word *innovate,* coming from the Latin *innovare* — *to
renew* — is defined by Webster as the act of introducing something new to make a change, especially in customs, manners,

and rites. It is the process by which an invention or an idea is translated into the economy for use. Innovation is usually distinguished from imitation in that it is a change new to the economy as well as to a particular institution. When the innovation is new only to an institution it is usually classed as an instance in the diffusion process, which is discussed in Chapter 3.

Although innovation is not limited to technological products and processes of the industrial and business world, this is the principal sense in which it is being discussed here. There are many fields where nontechnological innovations are of great importance — for instance, in social institutions and their interrelationships. Both invention (the conception of the idea) and innovation (the use of the idea) make up the total process by which new ideas are conceived, nurtured, developed, and finally introduced (1) into the economy as new products and processes, (2) into an organization to change its internal and external relationships, or (3) into a society to provide for its social needs and to adapt it to the world.

Boredom is often a potent psychic force behind innovation, because it generates a revulsion and then a search for alternatives — a process which is no small part of the dynamics of fashion, gaming, sports, exploration, and even some scientific investments. It's not a factor in the management world today, however; with the rapidly changing environment under the violent impacts of cultural and technological innovations, the manager is far from being a victim of boredom. Rather, it is this rapid change and threat to the survival of his enterprise that is his major psychic impetus to innovation.

The problem is that the usual organizational pattern is set up for yesterday's business, so that it becomes bureaucratic and inflexible, unwilling to open up and permit needed changes. Apparently only a crisis can transform the habitual organizational inertia into a readiness to try new tools and new ways; it has been said that wartime cooperation is never reached in peacetime. But the tempo of today's activities, the rise in competition, and the impact of technological innovation provide the crisis atmosphere. An institution can no longer afford the luxury of designing changes as does the viticulturist in the production of wines, where changes in wood as well as in wine are allowed to occur over a long period of time with little management of the process.

In his book about the approaches to stability and change which have been dominant in the history of human thought, Dr. Donald A. Schon depicts our present attitude as essentially one in which we conceive of our institutions as enduring.[3] We tend to see changes in values as deviance, undependability, and flightiness. We assume that values are firm and constant, and with some of us, maintenance of the establishment is even sacrosanct.

Technological innovation has become increasingly recognized as a principal, if not the principal, instrument in intercorporate competition, according to Dr. Schon. For individuals within the firm, technological innovation has become a major task and the main road to upward movement.

Schon looks first at invention. From a rational viewpoint, it is a goal-directed, orderly intellectual process. However, invention has a nonrational side in that it often works backward, and once the process of technical development begins, it does not usually move in a straight line but takes unexpected twists and turns. Need and technique determine each other, and answers frequently turn up in surprising places.

The rational view of innovation is that it can be managed and is subject to intellectual control. But this view ignores actual experience; a more relevant model consists of situations that are both unfamiliar and problematic. Action is required, but just what action is unclear. According to William James:

> All theories contain a penumbra of uncertainty, error and ambiguity which are irrelevant from the theory's point of view but become the basis for the next theoretical advance in the field.
>
> The great field for new discoveries is always the unclassified residuum. Round about the accredited and orderly

facts of every science there ever floats a sort of dust cloud of exceptional observations, of occurrences minute and irregular and seldom met with, which it always proves more easy to ignore than to attend to . . . anyone will renovate his science who will steadily look after the irregular phenomena.[4]

Uncertainty, therefore, is inherent in technical innovation. But as long as uncertainty remains, the corporation has difficulty functioning, because it is not designed for uncertainty or unclear objectives that do not permit measures of accomplishment. "The corporation cannot operate in uncertainty but it is beautifully equipped to handle risk . . . it is, from at least one point of view, precisely an organization designed to uncover, analyze, evaluate and operate on risks. . . . Accordingly, the innovative work of a corporation consists in converting uncertainty to risk."[5]

Innovation: Species of the Genus "Change"

About 200 years before the dawn of the Christian era, Marcus Aurelius said, "Observe always that everything is the result of a change . . . the universe is change." While we seem to be obsessed at this point in history with social, political, and technological changes, we can certainly learn to deal more effectively with them if we examine history, observe the process of change itself, analyze the environment we are facing, and study the basic concepts that are rooted in change — in particular, those changes which are innovative.

The basic ideas underlying the new "in" words such as *innovation, technological transfer, technological innovation, management of change, serendipity, synergism,* and so on are as old as the history of mankind. Webster makes a distinction between change and innovation, pointing out that innovation is the introduction of something new or something which differs from existing forms, and that it occurs as a result of initiative in planning.

In the educational field, innovation has been considered

a species of the genus "change." As of March 1970, some 2,500 innovative programs in education had been financed in the United States under Title III grants of the Elementary and Secondary Education Act (ESEA) passed by Congress in 1965 as part of the Great Society legislation. Compared with the $4.3 billion appropriated for Title I (benefits to low-income-family children), an allotment of $561 million (or 13 percent of the Title I figure) was devoted to Title III grants for innovation since fiscal 1966.[6]

Napoleon claimed that a general must change his tactics every ten years if he wishes to maintain his superiority. This practice may have sufficed for war in Napoleon's time, but with the rapid technological changes of our period, such a time frame is too long for strategic effectiveness. In a world turbulent with social, political, and economic changes, as well as the complex interactions of great moral, psychological, and cultural forces, change itself becomes commonplace. The need to adapt to successive changes — and in fact to manage certain of them — remains a major challenge to management. It is critical for an organization's survival, let alone its satisfactory growth and social contributions, whether these be profits or other values.

Thus we must first examine what Lebanese statesman Charles Malik calls "the mystique of change," and perhaps eliminate some of its arcanum. The scope will be confined to economic and scientific changes, although certainly political, educational, social, moral, and philosophical changes are interwoven in the management process for innovation.

Professor B. Lamar Johnson has suggested that the mystique of change might best be handled by the creation of a "vice-president in charge of heresy," while some waggish student has referred to such a position in the college field as "the innovative dervish" — a nonadministrative post.[7] Interesting changes are being made in the organizational architecture of nonbusiness institutions, and we as managers must watch these from an unbiased viewpoint to take advantage of their intellectual input. The discussion here, however, will be primarily concerned with business and industrial management.

—

In summary, our problem, at least in the Western world, seems to be that the environment is changing very rapidly and will continue to do so in the foreseeable future. It is essential to recognize this, understand the concepts of innovative change, and have the will to do something about it. Demands for change and innovation occur not only in business and industry but also in other fields: in education, with its student revolts; in government, with the mandate for change in foreign policy, housing, poverty programs, pollution, and so on; and in religion — for example, with the pressures on the Vatican for modernization.

Franklin D. Roosevelt said: "New ideas cannot be administered successfully by men with old ideas, for the first essential of doing a job well is the wish to see the job done at all."

Innovation: Social Dimension

A traitorous innovator, a foe to the public mind.
(Shakespeare, *Coriolanus*)

From podiums, in annual reports, and even through full-page newspaper advertisements, corporate presidents are voicing their plans and aspirations for their companies in combating industrial pollution, promoting urban renewal, responding to educational needs, and so on. Perhaps the movement is a bit late; nevertheless, it is significant. Management is assuming its duties and obligations to innovate in order to cope with the external diseconomies and social hazards that have arisen in our technically innovative society.

Today, security analysts who weigh a company's prospects consider the analysis incomplete unless a sociological input is included. Some trustees of nonprofit institutions are being urged by beneficiaries and constituents to channel funds into companies involved in eradicating poverty, pollution, and hard-core unemployment, and to liquidate stock in firms not actively engaged in socially acceptable purposes. While stock purchases may have been originally made on a financial basis, some trust officers are now being asked to justify their

choices on sociological grounds as well. This concern with
social goals formally started in May 1970, when a Boston syn-
agogue decided that it wanted a "peace portfolio." As a result
an enterprising analyst, Alice Tepper, created the Council on
Economic Priorities, a research firm to serve what *Business
Week* calls "the Dun and Bradstreet for the socially concerned." [8] Its first report, "Manufacturers of Anti-Personnel
Weapons," listed 105 companies and caused some investment shifts by socially conscious institutions. A fairly expensive subscription service, the Council on Economic Priorities
is one of many efforts to put a company's social policies on the
balance sheet. Another approach to the social dimension was
the January 1969 report to President Nixon by the Bell-Rivlin
Panel on Social Indicators under the U.S. Department of
Health, Education and Welfare. This report developed statistics for periodic stocktaking of the nation's social health.

The technological innovations which gave rise to the public's emphasis on social values have been well recognized.
Major opposition in the United States to such phenomena as
nuclear energy, urban development, and industrialization
focused public attention on the secondary consequences of
technological advances. This in turn has forced managers to
be concerned about what is happening in the world outside
of their own establishment.

These exogenous factors are acting to counter a trend
within management itself. As managements increasingly develop a systems approach, their tendency is to abdicate responsibility for the system per se, including its impact on the
environment, because they are preoccupied with the internal aspects of the system and with innovation in the management process itself. One way to correct this is to increase
the manager's awareness of the spectrum of the management
process and the total environment. The process of innovation
is only one small part of that broad spectrum. The hazard of
relying too much on models for guidance is a real one. Like a
man shaving in the mirror and becoming so absorbed in the
process that he lathers the mirror, model concepts can cause
management to mistake shadow for reality. The process of in-

novation must be experienced by the chief executive, not delegated or relegated to an abstract model.

Innovation: Organization and Guidelines

While logical thinking and reasoning can be applied to the process of innovation in step-by-step fashion, innovation, as has been pointed out, unfortunately is not always logical. Nor is it entirely cerebral; if we leave out emotion, we would miss the very essence of innovation. As Santayana said, "It is wisdom to believe the heart." Innovation is the area where the management scientist and the "manager-artist" can come together.

Any discussion of how to organize for innovation might well be called how *not* to organize for innovation. Since organizational dynamics are involved, there is no one correct way to organize for innovation after the chief executive has made a commitment to pursue an innovative course. The environmental situation, the experience, maturity, and open-mindedness of the organization, and many other factors will dictate what organizational design is most useful. While the behavioral scientists are planting some candles in the dark field of administrative science, most of their theories have not yet been reduced to practical guidelines that an executive can use. It is just a matter of time, however, until management practice catches up with some of the more advanced theories, which the reader will find in the references cited at the end of the book.

Because we know so little about human behavior, one of our major obstacles is securing acceptance of innovations in the fields of business, government, and education. Human nature resists changes, and we can learn from the sociologist, psychologist, and anthropologist something about the anatomy of this resistance (which will be discussed more fully in Chapter 3).

The key to the innovation process is the chief executive's role in understanding the innovation ethic and making a commitment toward innovation. The practical problem, however, then centers on the question of how to execute and sustain a successful innovative action. Ways of approaching this problem are described in Chapter 7. Max Frisch, the Swiss novelist, characterizes technology as "the art of organizing life so that we don't have to experience it." While innovation is like technology in that it is a technique without purpose and meaning of its own, innovation could be called the art of organizing life so that we do experience it. Innovation is a challenge to the organizing talents of the chief executive — who is usually inclined to delegate most tasks, who is uncomfortable in dealing with uncertainty, and whose "flight to the familiar" is a normal human tendency.

We have learned that computers, when given guidelines, can change their own programs and go through a learning process. It would appear that the program of innovation can also become a learning process within management, provided guidelines are established. However, setting guidelines for innovation requires considerable thought. We are so used to dealing with what is called public information — that is, the describable, measurable information which is logical and determined, as in the physical sciences — that we tend to neglect the private world of experience, of personal feelings, charisma, emotion, ideals, morals — things which are uncertain and ambiguous, which do not fit on the grid of reason.

The chief executive officer somehow must step out of his describable institutional life and get the flywheel of innovation started with a momentum of its own. Parameters must be established for projected new endeavors, but there must also be elbowroom for innovations of the serendipitous variety. No real guidelines can be set for these, other than the principle that the corporate attitude should be favorable to nonconformists and divergent thinking. Consider the baker in a small Florida town who was also a volunteer fireman. Twenty minutes after he had placed rolls in the oven, the fire alarm sounded. Rather than let the rolls burn, he took them out, thinking they were ruined. On his return three hours later, he rebaked them instead of throwing them out. Thus he discovered brown-and-serve rolls, for which General Mills paid him $400,000. The phrase "half-baked" doesn't have a bad connotation anymore.

Because of the obstacles to innovation often presented by large organizations, there has been a rash of small entrepreneurial companies formed and bankrolled by venture capitalists in the United States. Some of the rationale behind this "spin-off" of the entrepreneur from large companies is reviewed in Chapter 9 on venture capital. These small companies have been responsible for many major technological breakthroughs. The extent to which the same pattern can be followed in the areas of education and social innovation is yet to be seen. Some of the smaller schools and smaller government entities are innovating successfully in an experimental way and there is good reason to believe that this model will be increasingly effective in nonbusiness institutions as the innovation ethic becomes better understood and accepted.

The usefulness of innovation (meaning the practical application of an idea, as distinguished from the conception of an idea, which characterizes an invention) is perhaps best demonstrated today by the Americans and the Japanese. The process of changing ideas into money is understandably a fertile field for authors, consultants, professors, and various experts from almost every walk of professional life, as well as for promoters and some with dubious contributions. In 1950 about $24 billion was spent on research and development in the United States. Estimates for 1980 approximate $50 billion; yet some analysts predict that only 15 to 30 percent of this potentially useful technology will actually be put into application. Estimates for the present day are that only 10 to 20 percent of the potential from our existing computer technology is being used. The problem of diffusion and utilization of technology is one of communications, attitudes, organization, entrenched institutional momentum, public and governmental climate of opinion, and the nature of men themselves.

No matter how much a company spends on research and development, the yield on useful technology is very low. According to Du Pont, about 80 percent of its sales comes from products developed outside the company, despite its heavy research and development expenditures. One industry spokesman maintains that it is more important for a company to have a good "imitation and development" department than a good R&D one.

The challenge of the innovation ethic has even led to the creation of a new industry which the TTA Information Services Company calls the technology transfer industry.[9] As a vital element of this new service industry a TTA survey lists 86 companies in the United States whose major purpose is to supply venture capital for technology transfer. (The venture capital movement in the United States is analyzed and compared with that in Europe and other parts of the world in Chapter 9.) In addition to the venture capital element, seven other branches of the new service industry are identified by the survey:

> Patent marketing, by companies that market patents owned by themselves or others.
> Patent and product development, by specialized firms that invent professionally.
> International trade and licensing, by companies that specialize in the transfer of technology between countries.
> Venture development and management firms, which form new companies to exploit patents, processes, or technology by bringing together technology, management, and capital.
> Product search, by companies that locate potential new products for clients.
> Corporate search, by companies that specialize in merger and acquisition for corporate clients and that assist in negotiations at all phases.
> Specialized information services, furnished by companies that bring seekers and sources of technology together by subscription publications or custom research.

The TTA survey is one more organizational guide to the innovation ethic as it is fast evolving from its spawning ground in the United States. In reviewing this book the London *Econ-*

omist points out that the overall impact of the "newly discovered technology transfer industry" on American businesses is hard to evaluate, but that new products uncovered by the product search companies cost about 2 percent of what the client would have to pay for developing them.[10]

The impact of the innovation ethic will not depend so much on organization and guidelines as it will on the nature of man himself. In particular, it is the innovative entrepreneurs who can make both the innovative transfer of an idea to a use and the imitative transfer of an idea from one application to another. The imitative majority will eventually follow the innovating minority. It is up to leaders of institutions to develop an understanding of the continuous process of innovation, with its implicit "systole and diastole" of technological change requiring further social and technological changes.

2

Management Process Concepts- Developing a Setting for Innovation

One person procreates a thought, a second carries it to be baptized, a third begets children by it, a fourth visits it on its deathbed, and a fifth buries it. *G. V. von Lichtenberg the elder*

POINCARÉ was reported to have said, "Natural sciences talk about their results; the social sciences talk about their methods." And so it is with the management process, which is a combination social, political, and intellectual activity. General management theory and practice are relatively new in the scheme of things, and the entire process—from creation, discovery, development, growth, survival, innovation, and renewal phases of the *raison d'être* of the institution—is difficult to tie down to an adequate theoretical base. However, enough is known to characterize the entire management function for our purposes of discussion, so that the separate

16

processes of innovation and the management of change can be examined in the larger framework.

It is difficult to conceive of any managing activity, even the most rudimentary forms in the earliest societies, as having any intellectual basis except for a somewhat mystic and artistic use of natural leadership characteristics. While leadership consisted mainly of physical prowess in early times, it is now primarily an intellectual task.

The birth of science, with its techniques of synthetic inquiry, observations, and experimentation, is generally conceded to have taken place in Egypt perhaps around 5000 B.C. It seems to have been there that order first appeared as people learned to secure food through domestication of animals and plants, and as they learned to organize the work of the community into what we now call crafts, the foundations of the true sciences.

Men of art and science (who can be traced from the Ionian philosophers starting around 500 B.C.: Pythagoras, Hippocrates, Plato, Aristotle, Archimedes, and Hipparchus) brought us into the decline of the Roman Empire, through the Dark Ages and the revival of learning in the Renaissance, up to about the sixteenth century, the beginning of modern experimental times. In the process they did not develop the management discipline as we conceive of it today, because they did not have the need to consider the management function as a separate challenge. When productive capacity began to increase from the sixteenth century on, the problem of distribution of the surplus presented some of the first management problems. Today, with our complex social and organizational structures and highly productive societies, the skillful manager occupies an essential anthropological niche in our environment. The industry-oriented managerial class has its roots in our modern competitive culture rather than in the racial or hereditary traits of the ancient peoples who founded early science.

The cement that binds together the members of the managerial elite in our industrial society is composed of many tentative axioms of management: collegial resonance, behavior patterns, risk concepts, the managerial power structure, profit motivation, a survival syndrome, a managerial life style, and the social consciousness of the manager. The managerial cult, about which so much has been written in recent years, has an interface with both the artistic and the scientific worlds. The manager's interface with soft-science know-how must be as close as his interface with the scientific management disciplines now classified under an interdisciplinary format: mathematical modeling, measurements of risks in the decision-making process, organizational theory and design, experimental gaming, information control, a systems approach to organization, and innovative enterprise management.

A modern manager has a double-barreled task in utilizing the knowledge available from these two great areas. Considering the advances that are being made in both, he may well wish to modify the early seventeenth-century cry of Newton, "O physics! Preserve me from metaphysics!" to "O Mr. Chairman of the Board, preserve me from the behavioral and management scientists!"

The style of management varies with the varied stages of the enterprise as it goes from a small to a large organization, as its competition increases, as it moves from research and development into production and marketing, as it needs renewal, and as the rigidity of the establishment sets in. Managements are attempting to deal with the problem of rigidity in various ways: by creating separate organizational entities for dealing with the entrepreneur; by organizing according to functional or product-group concepts; by establishing advisory boards; by providing the classical functional lineup — manufacturing, marketing, research, engineering, accounting, and so on; and by considering the management activity in a systems framework. The methods chosen depend on the chief executive officer and the degree to which he is able to apply what might be called professional management. If he is production-minded, this will tend to be emphasized; if he has a marketing background, the same is true. A good executive, however, can rise above his functional heritage and develop a better-balanced viewpoint in directing the enterprise.

Intellectual Process

Three functions of management which have intellectual content are policy making, decision making, and control.[1] Conceptually, this constitutes a three-pronged intellectual attack on the management activity. But although management is basically an intellectual task, it also has an emotional aspect and a social aspect, and all three are distinctly different from the physical tasks the manager performs.

The intellectual management process might be characterized in the following manner.

It concerns growth. This is a stewardship obligation involving the formulation of objectives and the synthesizing of added values. It includes the establishment of a company charter; the development of a creed, a mission, an organization, and basic policies; and the creation of a corporate personality.

It is based on optimum use of resources and profits. To this end management must ensure overall standards of quality, balanced profitability, growth (both short- and long-term), and reproduction of the management function itself. The task also involves professional specialization in the subject of management and the development of alternate strategies, programs, and tasks.

It has certain interrelations. The intellectual management process is concerned with the interdependent areas of communications, control, conflict, duplication, personal attainment, and personal careers. It requires the sequential phasing of interrelated activities and the addition of value.

It involves the formulation of a practical philosophy of conduct. This entails setting objectives, planning, organizing, evaluating, and controlling—the five classical steps in management. It also includes the creation and development of new inputs, innovation, commercialization, spin-off, acquisition, renewal, coping with competition, and dealing with environmental factors. In the process management must establish time-limited targets which are measurable, as distinct from corporate objectives, which are timeless. Management reproduction is also a necessity that comes under this heading.

The complete management process consists of a group of sequential or, in some cases, parallel operations which constitute an economic activity as conducted on an industrial scale. To borrow a phrase from chemical engineering, management is concerned with each of the individual unit-process entities. Our discussion focuses on this continuum of unit operations and their relationship to each other, as well as to the outside environment.

Perhaps the fundamental laws of physical science, which underlie all technology, have a structure analogous to the management process. The three basic physical laws are (1) the conservation of matter and energy, (2) relationships pertaining to the equilibrium of the process, and (3) governance of the rate of change in systems which are not in equilibrium. While the physical sciences deal with material balances, heat balances, equilibrium relationships, rates of reaction, molecules, and gas laws, these unit steps are not unlike the sequential elements of the abstract and intellectual management process.

The continuum of management spans the period from 5000 B.C., when the Sumerians made the major managerial contribution of script and record keeping, to the present day. The Sumerian contribution was followed by the Egyptian recognition in 4000 B.C. of the need for planning, organizing, and controlling. According to some authorities, however, it was not until 1916 that the first complete theory of management was developed. At that time Henri Fayol presented his philosophy of the principles and functions of management and pointed out the need for teaching management in schools. Fayol built his theory around the process involved in management. He viewed management as a universal and practically the same process, regardless of whether its sphere of operation is governmental, industrial, or institutional. Because he saw management as a process, he approached it by analyzing the manager's functions of planning, organizing, staffing, directing, and controlling. Inasmuch as these functions depend on the individuals involved, his theory is somewhat eclectic. It recognizes pertinent aspects of the social sciences but does not actively incorporate them in its approach.

Although he was the originator of the process school of thought, Fayol's theory was not enthusiastically accepted or understood during his life. He was, in fact, a man whose concepts were ahead of his time by over 30 years. His orderly study of management from the board of directors down was discovered and appreciated in the United States only in the 1950s. Fayol was perhaps the first to define the functions of forecasting and planning, organizing, commanding, controlling, and coordinating, and his classical analysis has stood the test of time.[2]

Another way to approach the overall process of management is to examine the various stages of corporate maturity that exist. At least five are recognizable and are listed in Table 1; there may be more.

Analyzing his operation in terms of these stages of maturity may help the manager develop insight into the critical premises upon which to base the conduct of affairs, particularly when he is dealing with the management of change.

Political-Social Process

While a great deal of study has been given to the intellectual aspects of management, only limited attention has been devoted to the political (used in its best sense) and social aspects. Yet whether an organization is a business enterprise, a university, a government, or a religious order, its management involves certain common elements concerning the relationships between people, the organizational hierarchy,

TABLE 1

Management Stage	Some Attributes	Organizational Examples
Entrepreneurial	Creativeness Small size Great growth potential Uninhibited organizational structure At ease with uncertainty	Republic of Israel American Research & Development Corp. The new electronics companies RAND Corp. ADL, Inc.
Architectural and engineering	Rearrangement of resources to fit needs Great planning activity Preparing for coping with change Probabilistic assessment of risks	Republic of Mexico NASA Interstate highway programs Airport modernizations
Construction and rehabilitation	Regrouping of forces and assets Shakedown phase Changing management philosophy Growth may be slowed	MARCOR Federal Republic of Germany University of California
Institutional	Posture of a social trustee Custodial establishment Large diversified size Centralized administration Communications problems Resists change	GM Du Pont ITT Vatican's 1968 Curia DoD U.N. Harvard M.I.T.
Exploitation	Possesses some franchise Has advantage over competition Strong growth situation Management buildup underway Premium on innovative changes	Xerox Imperial Japan IBM Disneyland Kentucky Fried Chicken

and both the identified and the unidentified power structures. *The Peter Principle* tackles certain aspects of this in its humorous treatment of the "salutary science of Hierarchiology." [3]

It was Tacitus who said, "The secret of empire was out [*evolgato imperii arcano*]—an emperor could be made elsewhere than in Rome." The secret of empire as far as the management process is concerned is that the chief executive officer can be made otherwise than by board election. The key to the secret involves the political aspects of management. In the case of governmental institutions, during the past 25 years there have been successful coups in 70-odd countries—well over half the sovereign states in existence today. As a method of changing governments, the coup is much more widespread than normal election. In the case of business enterprise the coup is an unacceptable approach, except in the case of some well-publicized take-overs by shareowners or by invading corporate interests wishing to install a new chief executive.

The Power Structure

In attempts of the newer, far-seeing management people to take over from the established management in order to innovate, there may be some parallel in the efforts of the radical Right and the radical Left to usurp power from parliamentary regimes in the government area. The way in which such social movements have taken place is well described by historians and sociologists, who report how the electoral process can be set aside and control of a government seized by violence. A brilliant exposition was given in 1931 by Curzio Malaparte in his book, *Technique of the Coup d'Etat*. Malaparte's thesis was that in attacking or defending a government, it is futile to attack or defend the buildings or the ministries. These are merely symbols of power. The objective should be to paralyze and control the technological power points of the state—the rail and road centers, telecommunication networks, power stations, factories, and so on. Subsequent writers on political take-overs—for example, Edward Luttwak,[4] a naturalized British subject—have lived in many European countries and concentrate on political and military problems. Luttwak has not concerned himself with the powerful left- and right-wing movements but with the strategy of narrow elite groups who seize power from one another in underdeveloped countries, where the masses do not participate in the political process or have much influence on it.

However, the real challenge in modern management is not to displace the current management, but to coexist with it while attempts are made to introduce new philosophies, particularly in respect to innovation. This process actually means dealing with the power structure, whether the structure is overt or covert. The question is how to legitimatize an effort in innovative management which does not violate the establishment's way of operating under its own power structure. The enterprise as a whole must maintain the classical management theme and the traditional goal of producing more of the existing products with improved quality and lowered costs.

Although the need to introduce innovation does not justify a coup d'etat, some lessons can be learned by examining the dynamics of political coups. They always involve either a seizure of some power of decision within the present system, or the establishment of a supplementary or affiliated relationship to that system. The difficulty arises because those who are in charge of the system, whether it is a nation, state, or private enterprise, hesitate to relinquish any power at all.

Introducing an innovative group involves annexing a different style of ruling group to the classical one. The technique, unlike revolution, is not to confront, overwhelm, and smash the old order by superior power. On the contrary, as Luttwak says, it is a judo technique in which the system's own advantages in established weight are turned into weapons against it. When a small group of knowledgeable managers deals with a critical segment of the corporate apparatus which has been encouraged to innovate into another business requiring a different style of management, the group is always suspect and difficult to supervise. Given the benefit of new management philosophy and the artful use of organizational leverage, the group can establish itself successfully despite the fact that the existing power structure continues to operate effectively in its classical form.

Timing

However, a "renegade group" that establishes an innovative arm of the corporation is feasible only in certain corporate and environmental setups. The economic situation of the company must be ripe for accepting a new approach. The innovative group is in the best position when it is politically independent in fact, as well as in theory, of the parent organization's classical control. Yet it is very difficult for a treasurer, for example, to relinquish control, since his primary responsibility is to look after the assets of the corporation wherever they are.

In order to be successful the innovative management group must also have some type of political center. Its leader should be personally acceptable to and regularly encouraged by the chief executive officer and the members of the board, since his activities will probably be considered disruptive or threatening by the members of the establishment. Thus in dealing with a company which wants to innovate (or says it does), a key question turns on the company attitude toward fostering an innovative arm.

The modern corporation is almost invulnerable to direct assault by a group of young Turks who wish to set up an innovative wing. This, of course, is why so many entrepreneurs leave large companies and go out on their own. The only way that they could move a resisting establishment would be by some revolutionary approach to the decision makers, or by a take-over. Either tactic is unlikely to succeed in a corporate enterprise.

An influential small group may possibly accomplish a

more subtle coup d'etat by winning audience and favor with the management high command and then wooing subordinate management's endorsement. The subordinate levels, however, will usually reject any innovative threats. Efforts to innovate via an adjunct arrangement to an establishment may involve a long-time silent and patient effort—even with support from top management and the board.

Bureaucrat versus Innovator

The growth of corporate bureaucracy has implications which are crucial to the feasibility of establishing an innovative wing. First, side by side with the permanent machinery of the organization, its "political" leadership, a bureaucracy which has a structured hierarchy with definite chains of command, usually develops in most large organizations. There is a difference between the bureaucrat as an employee of the organization and as a personal supporter of the chief executive officer. If the bureaucrats are linked to the leadership, any display of power in the interests of innovation occurs in the form of a "palace revolution" which revolves around the manipulation of the chief executive and upsets the classical members of the establishment.

In the critical parts of the corporate apparatus, the treasury department, the accounting department, and the law department have inherent powers of repression which give discipline and rigidity to the basic power structure. The apparatus of these organizations is therefore to some extent a machine which will behave in a fairly predictable and automatic manner. An innovative group being established within a large corporation can take advantage of this machinelike behavior. During the introduction of an innovation, the group can define in advance the expected restraints and impedances and attempt to get resolution of policy or practice conflicts before they jeopardize the success of the innovation. The inherent resistance to change will always be there, of course.

It was once remarked that until the seventeenth century

the throne of Russia was neither hereditary nor elective but "occupative." A long series of abdications forced by the Boyar landlords and by the archers of the guard weakened the hereditary principle. Whoever took the throne became Tsar, and birth counted for little. Some contemporary management organizations have also come to the point of government by continual take-over. This has led to the decay of their corporate political structure, which is needed to produce a growing organization.

Strategy and Tactics

Were a chief executive officer to adopt strategy from the military or political arenas as the means of bringing his organization to accept the innovative process, he might try any one of the techniques of revolution, civil war, pronunciamento, putsch, liberation, insurgency, or coup d'etat. Probably the only practical, constructive strategy is the pronunciamento, which can be used by the chief executive to state the corporate objectives that are forcing change.

In some cases where there is internal conflict as to the aim of the corporation, an approach through insurgency may be acceptable. Rather than trying to seize power through the primary management, an innovative group attempts to set up a rival organizational structure, which can be separately based either politically or intellectually. Certainly the philosophy of managing this rival organization will be different from that of operating the parent, which is relatively immobile and resistant to change. For example, the Viet Cong in South Vietnam are aiming at setting up a new social structure and, incidentally, a new state.

A basic philosophical question is raised here with regard to the legitimacy of the management. Management's power is derived from the shareowners' endorsement of the board, which in turn appoints the management. If the old-guard management or board does not accept innovation, the "insurrection" or establishment of a rival innovative wing must stem from some tacit support by the shareowners or top management for the innovative approach. This is a very complex matter to get hold of in a corporate world, because it is difficult to determine realistically what the corporate will is. However, establishment of a rival innovative organization may in fact be in the basic interests of the shareowners, who see a gradual decline of existing business. In effect, they may favor change but be unable to articulate their judgment that some type of innovative effort is in the best interests of the corporation. It takes a courageous chief executive officer to innovate in a static or overly conservative establishment.

The prime example of an innovator working in a government-owned industrial complex was Enrico Mattei, head of the National Hydrocarbon Corporation (ENI). His activities have been described in almost James Bond terms in many periodicals. In addition to building up an industrial empire, he did battle with the international oil industry, became involved in Italian politics, forced foreign policy changes — and died under unexplained circumstances.

The Power Elite

In considering the political aspects of the management process, it is basic to recognize that all power is really in the hands of a small elite group at the top. This group is literate, educated, usually well-heeled, and secure. Therefore it is radically different from the majority of the employees, and it operates practically as a race apart. The average employee and the average shareowner recognize this, and they accept for the time being the elite's monopoly of power. Unless there is some unbearable situation which leads to an overthrow of management, most shareowners will tolerate conventional policies and pedestrian performance.

For the elite establishment, the "senior bureaucrats" in charge, the introduction of innovation presents a mixture of danger and opportunity. It is much safer to ride out crises and retire than to step forward as the sponsor of a new innovative regime. The average corporation does not endorse the risk required for an innovative thrust—the freedom to fail is not a right freely granted. If the elite top managers can remain aloof from a rival innovative wing, their established position remains relatively unthreatened. And it is much better to stay remote than declare opposition, since there is usually safety in inaction.

One of the great recourses of the establishment is literally to do nothing. This can be a form of "support" for an innovative department or subsidiary. In the centralized organization run by a narrow elite, power is a well-guarded treasure. On the other hand, power in the sophisticated, democratic, less intricately structured organization is free-flowing and very difficult to get hold of and direct. Generally, the introduction of an innovative organization in a highly developed, long-established company takes place only under exceptional circumstances. In an organization that is not so well developed, an innovative cadre of people has less of an establishment to perturb and is more likely to succeed.

One of the problems of an innovative group is getting sufficient corporate power to survive. If they want to make a success of an innovative department, the board of directors and their representatives must really and truly delegate to it the necessary power and resources. This requires realistic recognition of the hazards of innovation as part of the challenge of achieving the corporate goals. Corporate support must be spelled out, so that the innovative effort is not considered merely a perturbation in the system but is seen as a vital requirement for meeting long-term objectives.

Changing Nature of the Management Function

Our society is primarily a protective cocoon for the mechanisms of thought. *Dr. Donald Olding Webb*, Professor of Psychology, McGill University

Our managerial class is not often viewed as an intellectual subsociety, for it is so colored by its economic cocoon—

or rather by its hard economic shell—that its doctrines, dogma, and directives do not attract much scholarly attention. Managerial philosophies are changing rapidly, however. Making and selling products or services is no longer enough. More companies are perceiving their role as that of applying their technological and managerial know-how to satisfy society's total needs. Innovations geared to meeting social needs are no longer merely desirable; they are being regarded as conditions of survival. It is unfortunate that this awakening to social responsibilities is largely a response and a reaction to social pressures. Nevertheless, the turnaround in business attitudes is real.

Examples can be seen in the chemical industry. Dow is running a municipal waste treatment plant on a service basis. Du Pont has changed its widely recognized motto, "Better things for better living through chemistry," to "Ventures for better living." Monsanto recently ran a full-page advertisement in the *Wall Street Journal* that did not include chemicals in a listing of company products.

In his keynote address at the 1969 CIOS International Management Congress, Peter F. Drucker identified five new assumptions about management that correspond to today's realities.[5] Discussing the assumptions that formed the foundations of management theory and practice during the last half century, he has also challenged the tenet that entrepreneurship and innovation lie outside management's scope.[6] His view is that the focus on the bureaucratic side of management to the neglect of entrepreneurship reflected an economy undergoing a period of high technological continuity. This period required adaption rather than innovation, and the ability to do better rather than to do differently. Dr. Drucker predicts that in the future, entrepreneurial innovation will become the very heart of management and will be increasingly carried out in and by existing institutions. Moreover, the innovation thrust of the future will be on social as well as technological frontiers. Health care, for instance, is clearly as much of a challenge to the innovator today as the physical sciences were in the last century.

Innovation in the future will tap areas of knowledge other than science and will be channeled through existing businesses, if only because the tax laws in developed countries make existing business the center of capital accumulation. Modern innovation is capital-intensive, especially in the two crucial phases of development and marketing. The challenge, therefore, is to learn to make existing organizations capable of rapid and continuing innovation. They must reach out for change as an opportunity and must learn to resist continuity. During the fifty-odd years between 1860 and 1914, every major invention almost immediately ushered in a major new industry. On the average, these inventions appeared on the scene every two or three years.[7]

Management has been operating as a closed system, and the basic problem is to get managers to be comfortable in an open one. They must be ready for constant changes, not only in operations but even in management philosophy. The management concerned with innovation in the future should take a leaf from the book of the social scientists, since many management problems will involve social elements. The associational theory in social science, for example, describes built-in mechanisms to deal with the ongoing process of change. It analyzes the processes within the kind of social order that can change, yet maintain continuity with the past, in contrast to an order in which change is revolutionary and discontinuous.

An associational society is one in which people make membership choices among political parties, economic groups, religious organizations, and so on.[8] By implication, voluntary associations are organized to allow ultimate sovereignty to rest in the membership; decisions are made from the bottom up by the citizens, stockholders, or church members. The essence of the American social tradition is protest against unilateral structural control of any sphere of life against totalitarianism of the right or left. Management tradition, on the other hand, has tended to establish this unilateral control instead of allowing the structural independence that is necessary to nurture innovation. According to MacIver, such structural independence should be accepted as legitimate and it should be attended by the controversy and partisan alignments necessary in a heterogeneous society.[9]

The corporate society of the future will tend to be heterogeneous, and its management must have an entirely different philosophy and approach. MacIver says, for example, that in a social system we must accept differences as part of the "order of nature in social affairs. We accept political differences in social affairs and we must accept economic or religious differences in their affairs as a conflict theory of society." Management must learn to set up an organizational scheme which will serve as a principal means for the articulation and protection of differences. Managers must resist the monolithic order of the mainstream of business today and must learn to institutionalize "revolution" and make it gradual. This can be done by promoting free discourse in dealing with the strain and conflict involved in the shaping of management opinion.

In a free and open corporate society, diversity and partisan differences with respect to business judgment on innovations will be accepted as what might be called culturally legitimate. Farseeing sociologists even propose that multiplication of these disagreements contributes to stability of the overall enterprise. The point is that the approach to innovation in the future must accept the continuous-conflict way of doing business, since the innovative thrust is antithetical to the ordinary activities of the enterprise.

Of course, there is always a shortage of resources, and grass-roots responsibility for innovation cannot be meaningful unless the resources allocated to it are adequate. In the political field it has been necessary to transfer the decision-making power concerning resources right to the top in order to avoid starving innovative activities. One reason for the shift of functions from courthouse to state capitol to Washington, D.C., is the inadequacy of local resources for solution of local problems. The same is true of new business ventures which have limited sources of capital. Managers at the top must accept responsibility for allocating capital, time, and management interest to a new activity if they expect it to survive.

The Management Process

The eighteenth-century German physicist Georg Christoph Lichtenberg once wrote that there can be no greater impediment to progress in the sciences than the wish to detect its outcome at too early a stage. With this in mind, note that the governing board responsible for allocating resources to the creative effort needs to assume a risk-taking attitude, rather than the risk-avoidance or caretaker approach necessary at times in carrying on an established enterprise. Prejudices, biases, and experience accumulated in one style of managing cannot necessarily be transplanted to another. The creative group rejects structured directives as rapidly as the establishment builds up its "entreprenertia."

As one who organizes, assembles resources, manages, and assumes the risk of an enterprise, an entrepreneur is by definition endowed with managerial faculties at three levels:

The determinative or trustee role, in which the purpose, objectives, and overall philosophy of the enterprise are defined. If innovation is to be a characteristic of the enterprise, it must be sanctioned at the trustee-director-stewardship level.

The administrative or directive role, in which strategic planning and control of the assets is accomplished, general policies are formulated, and functional coordination is achieved.

The operating or executive function, which involves supervision, tactical planning, and control, along with local coordination and local policy.

Development and tolerance of entrepreneurial creativity, though it is required at all three levels, is seldom present in very good balance.

At the *determinative* level, the entrepreneur has his own personal value scale, philosophy, and style. Top management must understand that conventional wisdom is not to be imposed on him, since an entrepreneur is more comfortable in areas of uncertainty and has a higher tolerance for ambiguity than the classical manager. His opportunity to fail and his opportunity to obtain high rewards must both be accepted.

But the sponsor—that is, the investor or chief executive officer—has the right to demand a commitment from the entrepreneur with respect to the determinative-trustee responsibilities of his enterprise. This might be in the form of a clearly defined objective and timetable related to the potential market need of the new project, but not necessarily related to the market need of the sponsoring organization.

In turn, the entrepreneur has the right to know the rewards, if he is successful. These should be very different from rewards for success in a more secure and conventional enterprise. Equity participation is a common arrangement, but there are others in the form of perquisites, recognition, and so on. The entrepreneur should have the right to employ determinative input from an outside board of directors (or its equivalent), whose trustee role is primarily related to the objectives of the new enterprise and not to a risk-avoidance effort on behalf of the sponsoring organization.

At the *administrative* level, the management style should shift from a directive approach to a permissive one. Since permissiveness in a structured organization will be controversial and difficult to protect from efforts to erode it, organizational "weather stripping" of the entrepreneurial unit may be desirable. Isolating the unit from the rest of the business will minimize interaction with the highly structured organization and prevent imposition of the establishment's performance criteria and values.

The entrepreneur should have the right to obtain and pay for corporate services from any source, inside or outside the sponsoring organization. Corporate control, other than that required by statute, should be minimal. The entrepreneur needs freedom to organize, manage, and assume the risks of his enterprise. Within limits, he might be permitted to create his own style of total-asset management. Otherwise, if he unwittingly accepts the identity of the sponsor, he can be inhibited by the requirements of maintaining goodwill, prejudices, biases, and other intangible restraints which go along with the parent's corporate image.

Concerning the *executive* aspects of operating an entrepreneurial enterprise, the risk manager must not be told how to function tactically. He should be allowed to create his own mode of operation, which may be inconsistent with the style and tempo of the sponsor.

Dr. Herbert A. Simon has drawn an analogy between organ transplants and the rejection of innovation by existing corporate bodies.[10] In the case of computers, for example, businesses have vigorously resisted their use. It was only in 1950 that there was a serious business application, although computers had been in operation for over a quarter century. Businessmen characterized this invention as a "giant brain," a "thinking machine," or an "obedient robot." According to Dr. Simon, higher management has not yet fully accepted the computer for purposes other than routine clerical and accounting inventory chores, much less for help in management decision making. This innovation is difficult to assimilate because of the normal resistance to change. The real problem is how the innovative process can be grafted on to an established management process in order to enhance the growth and renewal objectives of the enterprise.

Dr. Michael Kami describes the mandate to modern management as

To provide corporate growth faster than the GNP, or die.

To deal with the change in external factors.

To deal with fast obsolescence in the economic life of the business.

To keep up with improving technology and competition, as well as coping with the tighter regulations, antitrust and otherwise, which are imposed on the business externally.

To acknowledge the factor of one-world shrinkage.[11]

Business is no longer isolated economically or geographically. The manager must deal with internal forces which are rising up to make his job difficult: the more complex communications due to organizational complexity; the general increase in education of management people, with more participative and consultative activities in the total management process; the increasing tendency toward bureaucracy, both in government and management; the shortage of talent necessary to conduct and expand the business; and the problem of motivating employees, with their changing sense of values.

Management's job of the future has a new element: It must provide for innovation of products and services which will insure profitable continuity of the enterprise. Thus the manager's former theme of more, faster, and smarter now has the intriguing fourth dimension of the innovation ethic — of being successfully different. Some approaches to this are discussed in Chapter 7.

Management Innovation

The McGraw-Hill Economics Department estimates that while in 1947 approximately 20 percent of all companies attempted business forecasts over three years or longer, 90 percent did so in 1966. The trend toward integration of technological long-range forecasting and planning, along with a simultaneous move from product- to function-oriented frameworks, is only now making itself felt. In the United States a six-year cycle appears to occur in the initiation of management innovation in these areas, according to Erich Jantsch: [12]

1953–1954	Corporate long-range planning introduced.
1959–1960	Technological forecasting introduced.
1965–1966	Integration of technological forecasting, and planning and orientation of functions.

A number of exceptionally well-managed companies have been ahead of these dates. Today, both long-range planning and technological forecasting are considered assets essential to a company's image. The next six-year cycle could well embrace innovative organization structuring, planning, and the management of ideas.[13]

Developments in Europe cannot be so clearly divided into cycles. Stanford Research Institute places the beginning of a substantial European interest in corporate long-range planning in 1964, which would indicate a ten-year cycle in management thinking. However, technological forecasting did start in many European countries at about the same time or even before the initiation of formal corporate long-range planning. As Jantsch says:

> Another major line of evolution in management concepts is now beginning to be influenced by technological forecasting. After a shift from horizontal to vertical organization — with marked decentralization into product lines (in the U.S. it is believed to be inevitable for companies with more than $400 million annual turnover) — the future trend toward function-oriented organizations involves the reestablishment of certain coordination which now takes the form of guidance toward future goals. Whereas the verticalization of American industry became almost complete in the twenty years following World War II, it is in full process in Europe. For European companies there seems to be an opportunity for a short-cut and a direct change from a horizontal product-oriented structure to a function-oriented structure.

In general, technological forecasting is much more closely related to corporate long-range planning than to R&D. D. B. Hertz says that in companies where no full structural integration of forecasting, planning, and research exists, actual technological forecasting or its coordination is often much closer to top management than is research.[14] His 1965 review showed that half the good research ideas in the chemical, electrical, and drug industries were originally suggested by top management. Churchill's warning that one should not attempt a major task from a subordinate position is obviously applied to technological forecasting in a number of advanced companies.

However, successful innovative management can be attained only partially through structures, organizational schemes, and special techniques. A fully integrated forecasting and planning system requires entrepreneurial qualities at each level and for each planning step. Thus it depends heavily on self-motivation in creative people at all levels. A well-known American company in the electronics sector decided to appoint as many general managers as possible. The term "general manager" did not signify a hierarchical position, but an employee who had access to literally all facts about the company and who was informed of all plans and policies. Before formal forecasting and planning were instituted, there were eight general managers in the firm. Five years later there were 150, and the goal was to have 1,000 general managers at all levels.

This calls for a certain greatness and profundity of vision on the part of top management, and the predominant attitude in industry is still far removed. The concept of 1,000 general managers can be contrasted with the example of the European company which was advanced in its thinking to the point of establishing a technological forecasting staff function. However, since knowledge of top management policies and plans for the future was considered a status symbol in this company, it withheld pertinent information from all persons except those in the highest ranks. Thus the technological forecasting group was not told about long-range corporate objectives; their planning for the future was either an extension of the present or pure guesswork.

Many companies of an innovating type, especially in the United States, have recently adopted the policy of stimulating ideas at all levels. Some follow through well; others do not. In United Aircraft, it has been found that ideas regularly emerge from groups in areas where company objectives are assessed systematically by the corporate staff and the results are fed back to the working level.

An ADL study of weapons system developments concludes that for 63 successful information-generating events, the following organizational styles and research managements were responsible: [15]

59 events	Adaptive environments.
3 events	Unable to define.
1 event	Only one man involved.
0 event	Authoritative environment.

In other words, an adaptive management environment is the one that stimulates innovation.

Greek legend has given us many useful words. Most managers are easily persuaded that their labors are *Herculean, Ixionic*, and *Sisyphean*. Ixion was bound to a constantly revolving wheel, and Sisyphus was condemned to roll a heavy stone uphill only to have it always roll down again. Without dwelling on the Herculean task of cleaning stables, we can observe that all three of these ancient "managers" were in difficulty most of the time. Such will probably be the case with managers forevermore. The changes that are taking place in the world, not only in economics but in all ideas, force the manager to manage himself as well as managing the processes of innovation and change. While he is changing as a person in his value system and interests, his job is also changing. And as Peter F. Drucker has said, "The most exciting thing about management is that there are no teachers, only fellow students."

It has been claimed that we are shifting from a society based on natural resources to one based on human resources, and that the most important ingredient of the new age will be ideas. To survive and grow, the manager must be able to stretch his mind beyond the management of physical resources and to conceptualize new philosophies of management. Professor Melvin Anshen of Columbia University's Graduate School of Business defines the problem as one of translating this vision and this change in management outlook into successful operations.[16] He suggests that there have been four stages in the transformation of the general management job:

First, the main task of the manager up until the past two decades was efficient administration of physical resources. The approach was short-range and unifunctional, with the dominant decision criteria being economic.

Second, beginning in the 1930s the concern with managing physical things was enlarged by a growing interest in managing people.

Third, after World War II the rapid growth in corporate

size shifted management's concentration from physical and human resources to finance. This management of money, although it was broader than the earlier focus on physical and human resources, "still has tunnel vision as far as dealing with tomorrow's problems and opportunities," according to Professor Anshen.

Fourth, the management of ideas is a concept broader than either management by objectives or long-range planning, which are only techniques. It goes beyond the concept of strategy, for "as there are alternate strategies for obtaining objectives, so there are alternate strategies for executing an idea that defines a central purpose of a business." Focusing on ideas contributes to realistic planning and to the development of more appropriate objectives and more relevant strategies.

While the changes in technology and concepts are moving rapidly inside the managerial sphere, the environment outside is moving even faster. Fortunately, additional management science techniques, as well as improved computers, are coming along to help managers cope with these matters. New organizational concepts are also evolving for dealing with the interface of the managerial sphere with the external environment.

Professor Anshen describes the opportunity that exists for a manager in terms of three options:

> (1) To mobilize all the company's resources around the concept of becoming a creative technological leader, viz, the first in the industry;
>
> (2) To organize resources around the central idea of becoming an early imitator and adapter of the successful innovations of the industry's creative leader; or
>
> (3) To become a low-priced mass producer of an established product, sacrificing high margins and high risks of innovation for the high volume and limited risk of low-priced imitation.[17]

A total scheme for operating a business can evolve from these options. The key is to have a single idea representing a top-management choice (among alternate strategies) for executing the scheme throughout the business. Another focus could be on ideas for conglomerates, and still another idea proposed by Professor Anshen as central to a business is the systems approach. This approach is valuable when a shift is warranted from defining the business in terms of a product line to defining it as delivering a complete system of customer values — for example, a shift from running an airline to providing transportation. Such a redefinition provokes the discovery of new businesses, such as environmental management, education as a life-span need, or "ghettonomics." New businesses can also be discovered by the abandonment of accepted ideas of industry boundaries, such as converting an asbestos company to a materials company, a petroleum company into an energy company, and a hospital into a complete cradle-to-grave health care institution.

To sum up, the concepts of the management process are broad enough to embrace continuance and renewal of business as usual, expansion of the present endeavors over a wider market area, and inclusion of an innovative dimension of growth, which in the long run competes for sovereignty in its own right. These new ideas can emerge only from a management environment that recognizes the anatomy of the management process and the changing management functions due to the changing environment, to competition, and to change in managers themselves.

3

The Innovation
Process

Inertia is not a detriment in every circum-
stance. When environments change slowly
inertia is a beneficial quality; it is a protection
against oversensitive response to fluctuation
in conditions. It is only when environments
change rapidly that inertia is a liability. *Sir
Eric Ashby*

THERE are few systematic studies of the innovation process.
While many articles and books deal with the subject of innovation from different viewpoints, including those of technology, history, psychology, and sociology, they do not take the
pragmatic approach necessary for the manager who is beset
by a host of operating problems and yet is consciously attempting to introduce innovation into his management process.
However, the manager can find good introductory background
to business innovation in two reports from U.S. government
sources.[1] One of these, a National Science Foundation study
by Sumner Myers and Donald G. Marquis, outlines the innovation types described in the following section.[2]

Innovation Types

Three distinct types of technological innovation are discernible in the real world, according to Meyers and Marquis. The first type concerns management of the technological change needed for very complex systems, such as communications networks, weapons systems, and moon missions. These major innovation systems may take years and millions of dollars to construct. They involve thorough long-range planning which assures that the requisite technology is available. Success usually turns on the skill of men who sort out the good from the bad approaches on a large scale. It is not yet a common type of innovation in most industrial firms, simply because few of them face the kind of systems problems that require it.

The second type of innovation is represented by the radical breakthrough in technology that changes the whole character of an industry. Jet engines, stereophonic sound, xerography, and the oxygen convertor are examples. These innovations are rare and unpredictable. They are predominantly the product of independent inventors or of research by firms outside the industry that is ultimately influenced by it, since the industries themselves are normally preoccupied with short-term research.

The third type, which Professor Marquis calls nuts-and-bolts innovation, is essential for the average firm's survival: "As long as your competitors do it, so you must do it. You must make technical changes on your own to get around the advance of the competition." This innovation is more intimately paced by economic factors than is innovation of the systems or breakthrough types. The garden-variety, within-the-firm kind of technological change, without which companies can and do perish, has certain characteristics derived from a complex mosaic of factors.

During the years 1963–1967, the National Planning Association conducted a study of actual innovation for NSF which covered 567 innovations in products or processes that had been made in the previous ten years. The 121 companies studied were drawn from five manufacturing industries: railroad manufacturers, railroad suppliers, housing suppliers, computer manufacturers, and computer suppliers. The innovations selected for analysis were judged by responsible executives as the most important in their companies.

Innovation, these investigators found, is not a single action but a total process of interrelated subprocesses. The garden variety of innovation may be carried out from conception to implementation within a single organization, but more commonly, it draws on contributions from other sources at different times and places.

According to the NSF study, successful innovation begins with a new idea which involves the *recognition* of both *technical feasibility and demand.* There exists a state of the art or current inventory of technical knowledge of which the innovator is more or less aware and on which his estimate of technical feasibility is based. At the same time, there is a current state of social and economic utilization which the innovator can recognize as an existing or potential demand.

The next stage is *idea formulation,* which consists of fusing the recognized demand and the recognized technical feasibility into the *design concept.* This truly is a creative act in which the association of both elements is essential. If the innovator concentrates on the technical advance alone, there may or may not be a demand for the innovation he produces. Similarly, if he merely searches for a way to meet a recognized demand, he may or may not be successful, depending on the technical feasibility of his project. Part of the idea formulation stage is the task of evaluation, which comes naturally after fusion of demand recognition and feasibility recognition into the design concept. Evaluation occurs all along the process.

Thus the innovator arrives at the design concept by identifying and formulating a problem worth committing resources to. This is followed by the *problem-solving stage.* Sometimes the information necessary for solution is already at hand in the state of the art, while at other times R&D and inventive activity are called for. Unanticipated problems usually arise along the way, and new solutions and tradeoffs are sought. In many instances the obstacles are so great that a solution cannot be found, and the work is terminated or deferred.

If the problem-solving activity is successful, a *solution,* often in the form of an invention, is found. This knowledge passes into the state of the art once patent protection is assured. Alternatively, the problem may be solved by the adoption of an invention or other input from the existing pool of technical art, in which case the ultimate technical change becomes simply an innovation by adoption or by imitation.

The *development stage* then begins. The innovator attempts to resolve uncertainties with respect to market demand and problems of scale-up. Innovation is never really achieved until the item is introduced into the market or into the production process, and until sales or the desired cost reduction are achieved.

The final stage named by the NSF study is the process whereby the solution is *utilized* and *diffused* in the marketplace. This stage is by no means always reached; only one or two products out of five achieve sales whose profits provide a breakeven return on the investment in the innovation.

Resistance to Innovation

First a new theory is attacked as absurd; then
it is admitted to be true, but obvious and in-
significant; finally it is seen to be so important
that its adversaries claim that they themselves
discovered it. *William James*

In the specialized world of psychological warfare, two kinds of power forces exist. One is an offensive force designed to reduce the enemy's belief in his own cause; the other is a defensive force designed to close ranks behind one's own cause and to reaffirm one's own ideas and purposes. The psychology of introducing change has some of these elements, for the establishment resists change and often closes ranks against the invader.

Professor James R. Bright of Harvard, in his landmark text-book *Research, Development and Technological Innovation*, lists 12 reasons why the establishment may resist a techno-logical innovation:

1. To protect social status or prerogative.
2. To protect an existing way of life.
3. To prevent devaluation of capital invested in an existing facility or in a supporting facility or service.
4. To prevent a reduction of livelihood because the innovation would devalue the knowledge or skill presently required.
5. To prevent the elimination of a job or profession.
6. To avoid expenditures such as the cost of replacing existing equipment, and of renovating and modifying existing sys-tems to accommodate or to compete with the innovation.
7. Because the innovation opposes social customs, fashions and tastes and the habits of everyday life.
8. Because the innovation conflicts with existing laws.
9. Because of rigidity inherent in large or bureaucratic organ-izations.
10. Because of personality, habit, fear, equilibrium between in-dividuals or institutions, status and similar social and psy-chological considerations.
11. Because of the tendency of organized groups to force con-formity.
12. Because of the reluctance of an individual or group to dis-turb the equilibrium of society or the business atmosphere.[3]

Professor Bright points out that the typical innovator takes little account of the possible responses of those who will feel the impact of his innovation, even though the same devices for resistance have been employed throughout history. These devices fall into classical areas of legal-political maneuvers, business and labor actions, propaganda and pressure-group techniques, and religious-moral positions.

Resistance to innovation is a central corporate problem, for the establishment is in a state of dynamic conservatism, striving for survival, stability, and continuity. The corpora-tion is modeled after the production process, meaning that it is rational, orderly, standardized, and predictable; but in-vention and innovation are both nonrational processes resisting control. As Dr. Donald A. Schon says, "They are precisely what cannot be managed and in spite of the best efforts of proponents of the rational view to harness them with mechanisms and to marshal them in orderly array, they keep popping out of the mold, dismaying those who attempt to control them even when they succeed." [4]

Resistance to change takes a number of forms: production slowdowns, tardiness, high turnover, tension, wildcat strikes, and arguments giving various pseudological reasons why the change will not work. The thesis of a benchmark paper by Paul A. Lawrence on this subject is that people do not resist technical change as such, and that much of the resistance which does occur is unnecessary. The five points made by Lawrence 16 years ago are still valid.

1. A solution which has become increasingly popular for dealing with resistance to change is to get the people involved to participate in making the change. This is not a good way to think about the problem and may lead to trouble.
2. The key to the problem is to understand the true nature of resistance. Employees resist, usually, social change, not technical change. The change in their human relationships usually accompanies technical change.
3. Resistance is usually created because of certain blind spots and attitudes staff specialists have as a result of their preoccupation with technical aspects of new ideas.
4. Management can take concrete steps to deal constructively with these staff attitudes. Steps include emphasizing new standards of performance for staff specialists, encouraging them to think in different ways as well as making use of the fact that signs of resistance can serve as a practical warning signal in directing and timing technological changes.
5. Top executives can also make their own efforts more effective at meetings of staff and operating groups where change is being discussed. They can do this by shifting their attention from the facts of schedules and technical details as to what the discussion of these items indicates in regard to developing resistance and receptiveness to change.[5]

Since innovation is something that corporate society must both espouse and resist, the establishment attempts to follow various strategies in order to cope with these two antithetical viewpoints. It compartmentalizes innovation, permitting it to occur in one part of the corporate structure, preventing it from affecting the other. It oscillates between support and resistance and confuses the corporate members by its on-again, off-again approach to change. It resists innovation while not appearing to do so by encouraging the development of ideas only to point out that none of them meet stringent criteria laid down in advance.

Cost of Innovation

In order to estimate the distribution of costs in successful product innovations and to examine the research component of the total R&D process, members of a U.S. Department of Commerce study group pooled their knowledge and attempted to discern a representative pattern.[6] The panel came up with the following rule-of-thumb figures.

In successful product innovations, the combined factors of research, advanced development, and basic invention account for only 5 to 10 percent of the total cost. Engineering and designing the product consume 10 to 20 percent. Tooling and manufacturing engineering (getting ready for manufacture) represent 40 to 60 percent. Manufacturing, including start-up expenses, involve 5 to 15 percent, while marketing start-up expenses range from 10 to 25 percent of the total cost of the innovation.

This breakdown indicates that the step we commonly call "research" accounts for less than 10 percent of the total innovative effort. The other components which we usually do not associate with innovation represent something like 90 percent of the cost. Managers in the technical industries are more or less aware of this, but it is generally not appreciated that the total cost of a successful innovation is so far beyond the costs of the invention and basic research stages. It is obvious, therefore, that R&D is by no means synonymous with innovation.

While the Department of Commerce study group focused

on successful product innovations, it attempted to get some idea of total R&D costs which include both successful and unsuccessful programs. As a rough measure, the total expenditures of U.S. manufacturing companies on R&D were compared with total net sales of these companies. In 1964 the total net sales were $293 billion, and the company-financed R&D expenditures were $5.7 billion. Thus the ratio of R&D costs to net sales was about 2 percent, a fact which further supports the conclusion that R&D costs are truly a small part of the total innovative effort in the manufacturing sector.

In the NSF study previously cited, incremental innovations which contributed significantly to commercial success were by and large not costly. Of the 567 innovations studied in 121 firms in five manufacturing industries, two out of three cost less than $100,000, and fully one-third cost less than $25,000 each.

Critical-Size Factor and Sources of Innovation

One sign of the need for careful study of the innovation process is the indiscriminate use of statistical aggregates purporting to show the comparative innovative performance of various countries. For example, R&D expenditures are often presented as a percent of gross national product. The Department of Commerce study concluded that such data are an inappropriate index of innovative performance. If such percentages were pertinent, then innovation would be as significant a factor in the nonmilitary, nonspace sectors of the United Kingdom (1.4 percent of GNP) and Belgium (1.5 percent) as it is in the United States (1.5 percent). However, since the annual national R&D budgets for these countries in 1961–1964 were $1.08 billion, $168 million, and $8.4 billion respectively, it is clear that they were not running a close race with respect to innovative successes. Such R&D data are obviously misleading as an index of innovative capability.

Further, it is fallacious to assume that when more money is spent on research and development (which, as mentioned in the previous section, is less than 10 percent of the total cost factor), there is an automatic multiplier effect on innovation in the marketplace. The main concern of the manager should be total cost and the total profitability of the entire innovative venture, which includes engineering, manufacturing, marketability, and marketing.

Many U.S. industries are apparently underspending on innovation. Although no hard facts are available to support this observation, the Department of Commerce study panel reviewed the "big sales" industries, and since there were no data on innovation expenditure, again compared R&D with net sales. In 1964 the steel industry with $17.8 billion in sales spent 0.6 percent of this figure on R&D. The transportation equipment industry with $34.3 billion in sales spent 2.5 percent on R&D, the chemical industry with $25.6 billion in sales spent 3.2 percent, and the drug industry with $5.0 billion in sales spent 4.5 percent.

Are the highly innovative industries progressive because of the way in which they respond to technological opportunities, or because their management meets the challenge of managing technological change? Is it that managements in other industries have not learned the importance of utilizing technological opportunities and innovative skills? Perhaps the answer is yes. In the United States this is a problem of education and not of antitrust, taxation, or capital availability, according to the panel's analysis.

But what about innovative performance as related to company size? Because of the lack of other data, the study group was again forced to compare R&D, not the total innovative process, with the number of employees. The panel found that less than 5 percent of the large companies perform almost 85 percent of the R&D, although this is not necessarily indicative of an innovative performance. It is important to distinguish between large and small sources of invention and innovation. The resources available are different, and not surprisingly, the riskiness of a venture and the manner in which it is undertaken are generally a function of the available resources.

Various studies have also been made of independent inventors, including inventor-entrepreneurs and small technologically based companies known to be responsible for a remarkable percentage of innovations and inventions—a much larger percentage than their relative investment would suggest. Some of the findings were as follows:

- In one study, Professor J. Jewkes and others showed that out of 61 important innovations in the twentieth century, over half stemmed from independent inventors and small firms.[7]
- A University of Maryland study found that during 1946–1955, over two-thirds of the major inventions resulted from the work of independent inventors in small companies.[8]
- A study at Harvard of 194 inventions in aluminum welding, fabricating techniques, and aluminum finishing showed that major producers accounted for only one of seven important inventions.[9]
- Another study at the University of Maryland of thirteen major innovations in the American steel industry revealed that four came from inventions in European companies, seven from independent inventors, and none from the American steel companies themselves.[10]
- An M.I.T. study of what were considered the seven major inventions in the refining and cracking of petroleum found that all seven were made by independent inventors. Contributions of the large companies were largely in the area of improvement inventions.[11]

Typical examples of important contributions in the twentieth century made by independent inventors and small companies are the development of xerography by Chester Carlson, catalytic cracking of petroleum by Eugene Houdry, Kodachrome by L. Mannes and L. Godowsky, Jr., automatic transmissions by H. F. Hobbs, the Polaroid camera by Edwin Land, titanium by W. J. Kroll, frequency modulation radio by Edwin Armstrong, cellophane by Jacques Brandenberger, and the cyclotron by Ernest O. Lawrence.

A great deal has been written about the massive approach by large organizations to technological problems and the resulting breakthroughs — witness the NASA program of mission-oriented research. No doubt these task-force efforts are indispensable to technological and economic progress. From a number of viewpoints, however, it is apparent that a unique cost-benefit opportunity exists in the provision of incentives aimed at encouraging independent inventors, inventor-entrepreneurs, and small technologically based businesses. The costs of special incentives to them are likely to be low and the benefits high.

Diffusion of Innovation

Americans spent $10 billion on R&D in 1960; 1971 outlay is forecast as $28.5 billion. In real dollars this increase is a different story. However, this tremendous research cost is an unrealized public investment until the resulting innovations are diffused to and adopted by the intended users. As a study by Everett M. Rogers demonstrates, it is a fact of life that a considerable time lag is required for an innovation to receive wide acceptance.[12] This is true despite the economic benefits of the innovations studied. For example:

- A 40-year time lag occurred between the first successful uses of the tunnel oven in the pottery industry and its general adoption.
- Over 14 years were required for hybrid seed corn to win complete acceptance in Iowa.
- About 50 years elapsed between the development of a new educational process and its adoption by all public schools. To put it another way, the average school lags 25 years behind the best practices. Similar time gaps have been found in other areas.
- The average U.S. farmer could support 50 other persons, rather than 27, if he adopted all the developed innovations.

Rogers' story of an innovation that failed — water boiling in the Peruvian town of Las Molinos — helps explain the resistance there can be to change. A major concern of the public health service in Peru was to introduce hygienic measures, such as teaching housewives to boil contaminated water. In a rural town of 200 families, a local hygiene worker who visited individual homes for two years was able to persuade only 11 housewives to boil their drinking water. The reasons for the failure can be traced largely to the cultural beliefs of the people, particularly to their customs concerning hot and cold foods. In their culture, hot food was considered appropriate only for persons who were ill; otherwise they ate cold foods. Since boiling makes water less cold, well people were prohibited by the cultural norms from drinking boiled water. Only the individuals who were least integrated with the culture could afford to defy the community norm on water boiling. Business managers as well as anthropologists can benefit from this illustration of the importance of cultural values, for the same type of cultural interfaces exist in the modern corporation.

As Rogers says, "An innovation is an idea perceived as new by the individual. It really matters little as far as human behavior is concerned whether or not an idea is objectively new as measured by the amount of time elapsed since its first use or discovery. It is the newness of the idea to the individual that determines his reaction to it." Technological innovations are new developments or combinations of the material, as distinguished from the nonmaterial, culture. Rogers points out that even in the case of technological innovations, the *idea* about the new material product is diffused in addition to the product itself. *Diffusion* is the process by which an innovation spreads from its source to its ultimate users. According to Rogers, the essence of diffusion is the human interaction in which one person communicates a new idea to another. Thus the amount of money spent on research, development, or application work is only one factor in diffusion. Coping with the barriers between cultures and accommodating separate value systems can be as important as the economic costs or economic benefits.[13]

Adoption is an intellectual process through which an individual passes from first hearing about an innovation to final adoption. It can have five stages: awareness, interest, evaluation, trial, and adoption. The adoption process differs from diffusion in that adoption is the acceptance of the new idea by one individual, while diffusion is the spread of new ideas in the social system, or the spread of innovations between societies.

Companies differ in the speed with which they adopt innovations. Four categories of adopters have been observed among industrial firms:

1. Innovators—the first firms to adopt a new idea.
2. Initiators—the firms who adopt the idea soon after the innovators.
3. Fabians—the firms who adopt the idea only after its utility is widely acknowledged in a particular industry.
4. Drones—the last firms to adopt new ideas.[14]

When the characteristics of individual employees are considered, undoubtedly some organizations contain adopters in all four categories. Over a period of time, however, a firm's record in dealing with innovative opportunities causes security analysts to classify it in one of the four categories.

One Cambridge, Massachusetts, nonprofit firm, the Organization for Social and Technical Innovation (OSTI), stresses the fact that the modern organization is built not around products but around business systems, and that the modern business firm bears a striking resemblance to organizations for social revolution.[15] The phrase "the diffusion of innovation" has been widely used in recent writings of anthropologists, economists, sociologists, and students of the history of technology, who see a social change as a process in which inventions—technological, social, or institutional—come to be realized as innovations and then spread throughout the broader society.

There is a stock of theories about the diffusion of innovation. According to Everett Rogers:

In a dominant theory of social transformation, innovation is spread out from a center. This concept is at the heart of theories proposed by economists and anthropologists alike to account for social change via the diffusion of innovation. . . .

Thus, at its most elemental level of conceptualization the diffusion process consists of (1) a new idea, (2) an individual (*a*) who knows about the innovation, and (3) an individual (*b*) who does not know about the innovation.[16]

In these new business system models, there is a shift in the principal problems of design from the design of products or techniques to the design of networks. As Warren Bennis has described, the individual's allegiance moves from membership in an organization or an agency to membership in a network or system as society matures.[17] The process of social learning broadens from successive sweeps on innovations through a society to the formation of more or less cohesive self-transforming learning systems in the business system movement.

Almost all theories concerning diffusion of innovation are somewhat blurred, for they are still in the formative stage. However, the traditional concepts of management theory and practice are undergoing vast changes. Tradition itself cannot be a creative force; it has always promoted formalization and repetition. In one sense, a tradition must constantly be destroyed in order to live. It must repudiate dead forms, and by receiving infusions of fresh energy, prevent living ones from becoming static. On the other hand, the new concepts and new ideas which are forces acting on the old establishment (the force of innovation, for example) must be restrained so that they are not destructive. Balancing and synthesizing the use of traditional and antitraditional forces is a stiff challenge to the manager in these times of rapid change.

A study by the Organization for Economic Cooperation and Development (OECD) gave considerable attention to the time factor in technological forecasting and the movement of a new idea, a new process, or a new product from its discovery through its creative phase, its substantiation phase, and its development and acceptance phase. One of the original studies by S. Colum Gilfillan concerned the 19 innovations voted most useful which were introduced between 1888 and 1930.[18] These showed the following time frame for diffusion:

Time from first thought of invention to first working model or patent	176 years
Time from first working model to practical use	24 years
Time from practical use to commercial success	14 years
Time from commercial success to important use	12 years

Another study by Gilfillan of more than 200 important inventions between 1787 and 1935 found that the time lag between first working model and commercial success was 33 to 38 years. In his study of these inventions Gilfillan observed that they come in "functionally equivalent groups," thus making effects easier to predict.

The 50 years which, according to the table, were required in 1888–1930 to go from working model or patent to important use compared unfavorably with the corresponding evaluations made of 200 of the 500 most important nonmilitary inventions introduced from 1785 to 1935. In this case, the average interval was 37 years. Of the 75 most important inventions introduced between 1900 and 1930, the average was 33 years. Another evaluation of 35 major innovations between 1911 and 1950 in the petroleum refining industry showed an arithmetic mean lag of 13.6 years between invention and commercial success. An analysis of past projects carried out by Lockheed found an average of 4.2 years from invention to innovation. A prime example of speedy diffusion was nylon, which took about 3 years from discovery to invention and 10 years more to innovation.

The time span between discovery and technological application or invention can be shortened considerably by tech-

nological forecasting, which reduces the lag between discovery and the start of the creative phase practically to zero.
For example, in 1945 Bell Telephone Laboratories formulated
new goals for peacetime research and decided to strengthen
solid-state work, with particular attention to its possible
contributions to communication technology for which available technology became insufficient. At that time the most
probable outcome expected by Dr. William B. Shockley, director of Bell's transistor physics department, was a solid-
state field-effect amplifier. After one year, it was clear that
only two avenues remained promising: semiconduction and
electroluminescence. Experiments carried out in 1947 and
1948 led to discovery of the point-contact transistor. Its importance was grasped at once, and the direction of the work
changed accordingly. Invention in 1951 and large-scale application developments started practically without delay. (Similarly, nuclear reactor development began in 1939 with little
delay after the discovery or demonstration of nuclear fission
in 1938.)

On the other hand, penicillin, one of the truly accidental
great discoveries, had a diffusion time lag of 10 years. This
was obviously due to lack of a clearly formulated goal and
normative direction. Sir Alexander Fleming accidentally
came upon penicillin in 1928, and it was not until René
Dubos described another antibiotic (gramicidin) in 1939 and
drew attention to the potential implications of the field that
a systematic investigation was launched with well-defined
aims. This lag contrasts with a time span of only one week
from the discovery of the maser principle to creation of the
first working device at Bell Telephone Laboratories which
will probably hold the record for some time to come.

The previously mentioned ADL study which evaluated
63 research exploratory and development events in the history of six complex weapons systems found that new weapons
systems depend upon many small inventions.[19] Only two
major inventions in this analysis made for the Department of
Defense contributed to the development of the systems. As
discussed earlier, it was also found that an adaptive environ-

ment, in contrast to an authoritative environment, is probably the absolute prerequisite for successful development. The following factors were discovered to be of most importance: (1) a clearly formulated need, (2) the availability of resources to be committed at once, and (3) an experienced body of people.

In summary, the time frame of an innovation's diffusion has changed considerably in this century. Before World War I there was a lag of about 30 years between the average technical discovery and its commercial application. Between the two world wars, it was about 17 years, and since World War II it has been reduced to 8 or 9 years.

In 1948 a transistor was developed, and by 1960 more than half the electronics industry had switched from vacuum tubes to transistors and was making delicate devices. The laser was invented in 1960, the Nobel Prize for its development was received in 1963, and by 1970 it had become a $3 billion industry. Today the interplay of discovery, invention, development, engineering scale-up, forecasting, entrepreneurship, risk, and uncertainty overlap and tend to blur together as innovations are created and diffused.

Examples of Innovative Organizations

As the general topic of innovation comes into the forefront for thinkers and doers, successful practitioners are being studied and extolled by journalists and academicians alike. This will be an arena of experimentation, exploration, and learning for some time to come. It is useful, therefore, to look at a few recent entrepreneurial-innovating firms in the United States which have achieved reputations in this fascinating field. It is companies such as the examples following, different though they are in size and situation, that are forging an innovation ethic. The first six examples are based on a *Dun's Review* article.[20]

Indian Head Corporation

In 1967 Chairman James Robinson and President Robert Lear decided that the Indian Head Corporation was going to be one of the top companies in the glass container industry. What made the decision so startling was the fact that Indian Head had no stake whatsoever in glass containers. Today it does: the firm successfully invaded an entirely new business. With 1970 total sales over $414 million, it already ranks sixth in the $1.5 billion U.S. industry, and glass containers account for 40 percent of the company's operating profits. Lear describes the management innovator this way:

> The truly creative manager has a strong fix on the ideal image and goals of his company. He has the knack for recognizing a major opportunity and the ability to put together the best possible combination of skills and resources to help exploit it. He is usually his own man with enough independence of spirit to allow him to think and operate in conventional as well as unconventional terms. Finally, he wisely supports himself with enough data and has enough convictions to let him challenge deep-rooted policies and company patterns if he believes his idea warrants it.

Lear planned glass acquisitions in four separate geographic areas. The next step was to acquire the best man in the glass business. He employed Earle Ingels, former president of Owens-Illinois' European operations. Sales for the group after acquisition started at $25 million, or 9 percent of the total, in 1967—and were $100 million, or 22 percent, in 1969.

General Mills

Of all the qualities the innovator must possess, perhaps the most essential is ironbound conviction. William K. Smith, vice-president (transportation) of General Mills, had this in his almost singlehanded war against the policies of the railroads in handling General Mills freight. The railroads' investment in freight cars for General Mills came to over $50 million. "The way they mismanaged our fleet alone," he snaps, "is indicative of why, with a total $8 billion inventory in freight cars, the railroad return is less than 3%." Smith was convinced that unless some far-reaching changes were made, inefficiency would continue unabated and distribution costs would keep mounting. He attacked the problem at two points—distribution control and freight car design. He installed the first computerized distribution control system and redesigned the boxcar, which had not been changed in the last half century. The new system keeps a perpetual inventory of all cars, and the new boxcar about halves the cost per unit. This was an example of an innovative management technique which was introduced in—or introduced in conflict with—an established management organization, planning, and control system.

Union Camp Corporation

Another innovator is William T. Bess, Jr., 47-year-old West Point graduate, who is executive vice-president of Union Camp Corporation. One of Bess's outstanding innovations is the "minimum investment plant," a radical new concept in expansion planning. Traditionally Union Camp, like other companies, constructed a plant based on maximum capacity, building ahead of the market. The size and capacity of Bess's minimum investment plant, however, is based strictly on regional market needs. This involved conceptual flexibility on the part of Union Camp—willingness to discard a traditional frame of reference for conduct of their regular business.

Pepsi-Cola

About five years ago, when head of the Pepsi-Cola Company's domestic beverage division, James B. Somerall (now chairman of PepsiCo Inc.) came up with a completely new concept in franchise arrangements which gave the franchise bottlers greater control over their own operations through consolidation. After two years of hard internal work, he sold the idea to Pepsi management. Phase I was co-op canning, in which Pepsi bottlers in a given area got together to make their own cans instead of buying them. They were able to save 5 to 8 cents a case in manufacturing material costs and as much again in costs of shipping and handling. Inventories were kept to a minimum. Phase II of Somerall's plan was consolidation by the bottlers in each market into one centrally controlled facility, which was jointly owned and operated. This was done through various arrangements: in some cases stock was issued; in others, a small corporation was formed by the franchisers; in others, an outside party assembled all the franchisers under a single ownership. Pepsi merely acted as a catalyst in putting the consolidation together. This innovative approach has changed the traditional relationships between major factors in the beverage business.

Singer Sewing Machine Company

At the Singer Sewing Machine Company, a subsidiary of Singer Company, 44-year-old Alfred di Scipio became vice-president of the North Atlantic consumer products group a few years ago and attacked the problems of a staid old-line company on many fronts. He revamped the management staff and launched a five-year expansion program budgeted at $66 million—more than Singer had spent in the previous fifteen years. Nearly all the 2,000 Singer retail outlets in the United States were renovated or relocated, and the same program is being applied in Singer's 1,400 European outlets. This is an example of innovative renewal of a corporate strategy on a worldwide basis.

Dictaphone Company

Under President Walter W. Finke, who took over several years ago, Dictaphone has become one of the outstanding innovative companies in the industry. A recent check showed that 52 of the company's 60-man executive group were either new to the company or had been upgraded, and that sales had risen nearly 25 percent. Finke's secret, according to one of his admirers, is his Pied Piper-like magnetism. People come to him out of the blue, and they tell their friends. Finke gives much closer attention to hiring management talent than do most executives. "Our organization is built around people, not the organization chart," he says. Ability to attract and keep entrepreneurs is a necessary attribute in the chief executive officer of an organization that expects to be innovative.

Norton Research Corporation

In 1939 Richard S. Morse was a young M.I.T. physicist working for Eastman Kodak in a subsidiary called Distillation Products, Inc., formed by Kodak and General Mills to pursue their interest in vacuums as a possible technique for drying film.[21] In 1940 Morse resigned and went to Boston to form a company built around the commercialization of research and the development of vacuum technology. The history of his innovative venture illustrates the contributions and difficulties of an entreprenuer, the tribulations of launching new enterprises, and the necessity for changing organizational concepts as innovations become institutionalized in their own right.

His new company, the National Research Corporation (later to become the Norton Research Corporation), was formed with some Boston venture capital men — among them William Coolidge, Charles Francis Adams (now chairman of Raytheon), and William M. Rand (later President of Monsanto, now retired). Each pledged about $5,000 for a total $46,000 advance to the new enterprise. NRC turned to military work for a first project: it obtained some licenses on patents from M.I.T. to use lens-coating techniques and sold the work profitably to the Army and Navy. The Army needed magnesium both for aircraft parts and as an incendiary, and NRC researchers worked out a way to scale up the Pigeon ferrosilicon process. This led them to invent a novel diffusion vacuum pump, which found its way into the Manhattan Project.

In 1943 Morse decided that the most potent area for commercialization of vacuum dehydration lay in developing a
process for powdering orange juice. He succeeded after much
difficulty, but none of the major food processors displayed
interest. However, Morse was able to attract the attention of
the First Boston Corporation and the investment banking
firm of Payne, Webber, Jackson and Curtis. In June 1944 NRC
went public in one of the earliest successful solicitations of
venture capital from the public for a purely technical enterprise (this is described further in Chapter 9).

With the capital, NRC built a pilot plant in Florida and
interested the Army Quartermaster Corps in the product.
This led to a contract for half a million pounds of orange juice
powder, to be delivered in April 1946 for a price of $750,000.
Unluckily, the powder's taste went off in temperatures above
70 degrees, so that it was not practical for storage in the South
Pacific. NRC was unable to resolve this problem in time;
however, when the war ended it again turned its efforts to
the domestic market and to the yet unsolved difficulty of
taste degradation. The eventual orange juice product has
since become Minutemaid — produced by a subsidiary corporation, Florida Foods, which is now publicly owned. Unfortunately, some of the market aspects had been neglected,
so that NRC had difficulty in getting the product going and
financial reorganization became necessary. This experience
demonstrated that it is not enough to organize a superlative
team of scientists and engineers and expect the markets to
materialize. It also became clear that even if the market is
there, the creative people who dream up the products and
processes to serve it are not necessarily the best ones to supervise the transition. The model of the organization must be
changed as the enterprise moves from start-up to commercialization.

The decade 1949–1959 was a fertile one for NRC, with
new technical ideas and some modest ventures. Holiday Coffee, a precursor of today's freeze-dried coffee, fell victim to
the tight money market of the Truman years. Fiber Research
Corporation was formed to capitalize on a novel device for
precisely grading wool, but it floundered because the wool

industry concluded that it did not need precision. A whole series of projects in vacuum technology, petrochemicals, and special alloys represented technical ideas far ahead of their time.

After the launching of Sputnik in 1957, Morse was asked to come to Washington, where he stayed under Presidents Eisenhower and Kennedy and, of course, under Defense Secretary McNamara. Hugh Ferguson, formerly vice-president of Grace Chemical Company, took over NRC during this period. Ferguson was strong in financial control and got it back on an even keel. In 1963 National Research merged with the Norton Company, and the operating divisions of NRC continue to function within Norton. The NRL Laboratory became the Norton Research Corporation. Morse returned to Boston in 1962, has taken on several directorships, and is also a part-time advisor in Washington. He is now a senior lecturer at M.I.T.'s Sloan School of Management and devotes some of his money and a lot of his time to new enterprises in the Boston area.

3M Company

3M is considered one of the best-managed industrial companies in the United States, in the sense that it has compiled the most remarkable growth record over the past 30 years. During that period sales increased 120 times — from $14 million in 1938 to $1.7 billion in 1970. Sales and earnings rose at rates of 16.6 percent and 13.6 percent respectively. In market value 3M grew roughly 18 percent per year since 1938. Approximately 25 percent of its 1969 sales were in products manufactured in the previous five years. 3M's market philosophy is "Look for uninhabited markets." Its success has come in large part from an ability to develop entrepreneurs from within.[22]

The technical style has been characterized by Robert M. Adams, R&D vice-president, as "Start little and build." 3M wants the men in the laboratory, in marketing, and in manufacturing to bring in new product ideas. It prefers to have these ideas come from the bottom up rather than the top down. The carrot in front of each budding entrepreneur is the possibility of becoming general manager of a new business, of working up such a hot project that management has to become involved whether it wants to or not.

Three financial criteria used by 3M are:

1. A 20 to 25 percent pretax return on sales.
2. A 20 to 25 percent return on stockholders' investment.
3. An earnings growth of 10 to 15 percent per year.

To meet these average requirements, considerable attention must be paid to the new ventures to be sure that as many as possible flourish in an above-average fashion. 3M uses the project manager scheme, putting one man in charge of the project in the early stages and letting him follow through. On choosing project managers, Adams comments:

> The problem is . . . you have three sorts of men, inventors, entrepreneurs and businessmen. They are not necessarily the same person, though each of them thinks he does the other's job fairly well. Before we choose a project manager we look at a man in terms of how well he will ultimately run the business. We prefer not to make a project manager lightly. If a project fails, the project manager has lost some time in his career in the attempt to become a general manager, but he really does not get hurt in the process. People who take on projects generally accept the possible penalty of losing time in getting to the top.

Market analysis at 3M goes hand in hand with the evolution of the product idea. To gain financial underpinnings for his project, the 3M entrepreneur must convince management that his particular product or service idea is a better one than the other candidates at the time. The product or service champion at 3M turns out to be one-quarter technical man and three-quarters entrepreneur. President Harry Heltzer puts the credo engagingly:

> We're in the business of gambling on individuals. If a man has been a pretty good judge in the past of what he has said he is going to do and has done it, then if he comes in

and says, "I want to start importing moon dust," I guess I'm likely to let him try. In a like manner, if a fellow who has been around here a great many years comes forward and says, "This year we've got a tremendous breakthrough that's going to cause all sorts of things to happen," and he's had four or five years in a row of not delivering what he said, well, I'm likely to give him an argument.

This implies more of a one-to-one link between the man with an idea and the chief executive officer than is really possible in a company the size of 3M, but it echoes the theme that ran through Heltzer's own experience years ago when he was struggling to make 3M's Scotchlite a viable product.

There is, however, a bedrock principle of free communication within the company. The motto is: If you need help, go find it anywhere. This, of course, requires two corollary principles: Know where to look for the information you need, and have it be made available when you ask. Many people have put their finger on these rules as one of 3M's key strengths over the years. The emphasis on internal communications is seen in the well-known 3M Technical Forum, which the company established to encourage professional people to mix and exchange ideas. This is an internal technical society, run by the professionals, which holds seminars on all sorts of technical topics drawn from the work going on in the various company laboratories. It has been broadened recently to include management and behavioral science subjects. Once a year the Technical Forum members are shown the new products just about to be introduced in the marketplace. The result of these activities is not only intense communication on who is doing what, but a stimulating race to see who can do it best.

Another company innovation, the New Ventures Division, is in charge of the care and feeding of internally generated business ideas and the development of a more integrated marketing strategy. This division also examines the possibilities of growth by acquisition. 3M has a low turnover of technical employees, reported as 2 to 3 percent because the work is exciting and profitable and because the company requires each professional to sign an exceptionally tough employee agreement. In essence, the agreement prevents a man from working for a competitor for two years. Since 3M is so diversified, it can be a long dry spell.

Texas Instruments

Texas Instruments is well known throughout the world for its fabulous annual compound rate of growth and earnings per common share, both of which are well over 25 percent. TI attributes this triumph of management to its systems for planning and control of the business, present and future. To be totally effective, TI believes, planning and control must be a continuous process—an established way of life and an everyday working tool. The principal responsibility of the corporation's board of directors is to ensure corporate self-renewal, a process based on three principles which control the organizational structure and purpose.

The first is that in our complex world, which was brought into being by the expanding technological revolution, technologically based companies such as Texas Instruments exist to create, make, and market useful products and services to satisfy the needs of its customers throughout the world.

The second principle is that the opportunity to make a profit is TI's incentive to create, make, and market useful products and services. Texas Instruments is convinced that corporate renewal begins with innovation, which is the key both to useful products and services and to profitability. The chairman, P. E. Haggerty, stated in an address on corporate self-renewal (June 12, 1967):

> First I must make clear the breadth of innovation we have in mind. Too often in this technological age we associate innovation automatically with research and development based on the physical sciences. But the fact is that critical innovation—regardless of the field—may occur in the

make and *market* functions as well as the *create* function. Further, the effective innovation is the integral (that is, the sum total) of the innovation in all three of the categories: create, make, and market. A company which is outstanding in research and development and truly creates useful products and services, but which is relatively weak in innovation, in making and marketing, may well be outperformed both in terms of product and service contributions and profitability by another organization less outstanding in product and service research and development but better balanced with a high level of innovation at all three of the categories: create, make and market.

The third principle is that the process of creating change at Texas Instruments is one of deliberate innovation in the create, make, and market functions — and of managing this innovation to provide continuing stimulus to the company's growth in usefulness to society. TI has a formal system of describing in writing, succinctly and completely, the strategies to be followed for growth and development throughout the company.

Tactical action programs (TAP) are then developed to implement these formal strategies. In 1967 the company had a total of 77 strategies and 591 tactical action programs. Of the strategies, 23 were identified as having a potentially major impact on the corporation; that is, if they were successful, they would generate at least $50 million a year in net sales billed over a period of at least five years. TI considers its OST system (objectives-strategies-tactics) an essential part of its ability to manage change innovatively and its fundamental tool for assuring corporate renewal.

Texas Instruments grew at an average compound rate of 30 percent per year from just over $2 million in 1946 to $828 million in 1970. Earnings after taxes have followed a similar, if somewhat more erratic, curve, growing at a slightly higher compound rate of 31 percent per year since 1946. Certainly TI's sophisticated management organization and style is a worthwhile case study for those seeking new concepts in dealing with change and innovation.

The Bell System

In the Bell System, innovation means renewal: improvement of the old industrial techniques or development of new ones. Innovation implies growth in the capability of technology to meet more challenging needs and opportunities in a rapidly changing world. Jack A. Morton, vice-president of Bell Telephone Laboratories, is a frequent writer and lecturer on Bell's innovative concepts. It is his thesis that technological innovation depends upon the close and timely coupling of industrial goals with relevant research.[23] The process is one of generation, perception, and conversion — conversion of basic research into technology for a needed industrial goal, and conversion of this technology into goods and services people need and are willing to pay for. Innovation is thus a total process, all parts of which are related and necessary to the purpose. Morton says:

> It is not just a flash of inventive genius, not just the discovery of a new phenomenon or understanding, not the development of a new product or manufacturing technique — nor the creation of a new market. Rather, the process is all of these things, each above some critical threshold, acting together in an integrated way towards an overall goal. All of the parts and the whole of the process depend on creative, cooperative actions of people — it is "people process" with a purpose.

Morton stresses the fact that management and creative people should not be in basic conflict, that they have a common bond in relevant technological innovation:

> One needs to get it, the other needs to give it. Accepting this, what can be done to make relevant innovation challenging and rewarding? Can we make innovation to a purpose, self-motivating to the individual? In the systems approach to the innovation process, management's role in the people process of innovation can be likened to that of a systems engineer for a complex information processing machine.

The Bell Labs functions are three: basic research, applied research, and development and design. These activities are coupled with each other and with Western Electric's functions, which entail the provision of improved communications, hardware, and software to associated companies. Western Electric in turn has three components: engineering for manufacture, manufacture itself, and distribution and installation. Utilizing Bell Labs' general scientific knowledge is the firm's primary purpose.

The innovative process, which depends vitally on communication between people, involves language, space, organizational structure, and motivation. In the Bell System, the recognition of the coupling principle has brought about some interesting organizational and spatial changes. For example, development design people were moved into Western Electric premises, so that the spatial barrier between the organizations was replaced by a spatial bond. Organizationally these professionals "belong" to Bell Labs, but spatially they are strongly linked to the Western Electric people. The applied research and development design functions were integrated organizationally within Bell Labs at the lowest level that would be consistent with group size and common technology.

Jack Morton has summarized his thesis for Bell Labs as follows:

> Technological innovation is not a single function or a random act . . . it is a total process with specialized but connected parts, all responding in some coordinated way to overall system goals. As a process it can be studied and managed from a systems viewpoint. A total innovative process is performed by people with widely differing skills and motivation. As a "people process with a purpose" it is vitally dependent upon the communications barriers and bonds between its specialists, and on the unifying challenges and rewards provided by the system's goals.

General Instruments Corporation

General Instruments Corporation has worked in competition with Bell Labs in the semiconductor technology field since 1951. Its management has learned from experience that it originally made a mistake in R&D by looking upon innovation as something which perhaps requires much planning and much systems orientation, but very little directive management. General Instruments has found over the last 19 years that the manager of innovative efforts must, in fact, overtly manage. The mistakes that the engineers and scientists make is that they remain thing-oriented. Once a system, organization, or environment has been designed, they expect it to work as smoothly as a functioning machine. The difference, as both General Instruments and Bell Labs emphasize, is that an innovation manager is not dealing with things; he is dealing with people, with constantly changing requirements, and with the very abstract concept of interpersonal communications. And this is a full-time management job.

General Electric

GE's investment subsidiary, Business Development Services, Inc., was formed in 1969 as a vehicle for this multibillion-dollar parent to branch out into new business areas. Transcending all the problems that a large firm has with antitrust, tax, accounting, bureaucratic conventionality, and institutional momentum was the people side of the equation. True entrepreneurs become ever less enthralled with a large corporate environment; they wish to start their own companies and be their own bosses. As a result, the games play or games plan of General Electric created this new subsidiary to nurture new ventures.

By late 1970, Business Development Services had invested over $5 million in seven small companies and is continuing its plan to invest several million more in other new ventures and support of the earlier investments. Its present thinking is to give these new ventures ample time to make good. Investments run between 20 and 45 percent of each company, all of which are privately held. GE is also considering licensing unused GE patents and providing cash for new companies that are in fields related to the patent estates.

GE hopes to support these minority investments until they can either be merged into an appropriate part of the parent corporation or go public if the majority of shareowners wish it.

4

Planning
Innovation

What use is a newborn baby? *Anonymous*

Every growing enterprise faces strange new forces and counterforces which not only impinge on it from the outside, but also from within.[1] The external forces, many of which have yet to be properly recognized, are socioeconomic, cultural, and political. The internal stresses and strains, including entrepreneurial thrusts, personnel conflicts, antiestablishment pressures, and innovative forces, are the subject of increasing study by behavioral scientists. Management know-how which can deal with the alien world outside and the complex world inside can mean success or failure to an institution. The task deserves the finest management talents, tools, and techniques available.

Strategic planning and intelligent collaboration are key elements in normal growth or innovative development programs, but the objectives of the respective efforts must be clearly set forth. Planning is inextricably involved with objectives, goals, environmental factors, organization, controls, renewal and innovative efforts, personnel, value systems, and factors that are cultural, intellectual, entrepreneurial, social, and political. As such, the planning function is part of a total

73

directive management process in which decisions are made and responsibility is taken for those decisions. The permissive character of any voluntary or laissez-faire effort, say within a creative cadre of researchers or entrepreneurs, should not overshadow the need for directive characteristics in the overall management of the organization. Amateurism in administration, allegedly tolerated in some sectors in order to "focus better on basic issues and professional satisfaction," creates a vacuum of weak management which reduces the effectiveness of the entire system effort.

Masterful administration of the unforeseen is the duty of the manager, whose stock-in-trade is decision making. He must maintain a consistently good batting average on decisions arrived at within a planned frame of reference despite a surrounding of uncertainty. A successful decision maker matches his techniques to the particular situations he faces as he pursues his long-range plan.

- Puzzle situations are those in which he can put together all elements and have the facts sit still for him.
- Risk situations occur when some of the facts or consequences can be assessed as probabilities, and the manager is at the mercy of calculated risks.
- Uncertain situations arise when there is not enough information to probabilistically assess the risks; progress involves a leap into the unknown to get this information.
- Strategies are useful when the decisions in any situation involve other people who have goals and hopes which may be different from the manager's. Esthetic and moral problems arise when he must fit his behavior to rules which are not a part of his own value system, and when he tries to do what's right.
- Tactics are the means by which the strategies are carried out.

In a puzzle situation—for example, one where the task is to design a bridge or locate a plant site—the puzzle solver deals with facts rather than people. Risk takers assign probabilities to future events on the basis of what facts are known, since time or resources are not available to do complete research. Balanced judgment is important in selecting the significant variables; as the Russian proverb goes, "Measure seven times but decide only once."

Strategic decision makers combine puzzle solving, risk taking, and acting in situations of uncertainty with the ability to anticipate how others will react to a decision. Since people are involved, good strategists are usually people-oriented as well as good controllers of their own behavior.

Developing esthetic and moral judgment requires long training that exposes the manager to the fitness of things. He must have an instinctive knowledge of what is right and what is wrong, for this will be his guide when he reaches the decision point. A sense of values is needed as much in business as it is in private life. One anthropologist has said that the two most productive areas for investigation of the history of man are his garbage dumps and his graveyards. In the first we find what our ancestors discarded as useless; in the other, their choicest and most characteristic offerings.

Innovative ventures are particularly troublesome in that they involve a preponderance of uncertainty rather than calculable risks. Hence strategic decisions for innovative ventures must combine aspects of the puzzle, the risk, and the leap into the unknown. It is here that a sense of values, experience, intuition, and mental flexibility are important.

Governments in the developed world now recognize that they cannot cope unilaterally with such contemporary problems as air and water pollution, housing and urban development, mass transit, racial unrest, and deepening social difficulties. Indeed, no single entity can manage these problems. They demand the best innovative minds of our generation from all fields, as well as effective collaboration and planning. Only a task force bringing together the most imaginative and realistic approaches taken from education, government, business, and philanthropic institutions can begin to grapple with the issues of the day.

In the developing world there is the additional challenge

that governments (and some voluntary nonprofit organizations seeking to aid these developing nations) must be educated to accept the profit ethic alongside the service ethic as a long-term measure of accomplishment. No nation and no people want to exist permanently on the dole. The key to a nation's enduring economic and social health is found in the dynamic words service and profit.

In the case of philanthropic organizations, their effective and planned use of resources (which is necessary to preserve the support of their contributors) is analogous to profitable performance in the business world. Further, they can collaborate with each other outside the network of antitrust provisions and other restrictions which sometimes interfere with cooperation between profit-oriented groups.

Whatever the type of organization, professional management will be called upon increasingly in the future to provide innovative solutions to local, national, and international problems. To meet these demands, managers will need to utilize the latest techniques and concepts of strategic planning.

What Is Planning?

Planning involves establishing the organization's objectives and purposes, setting policy guidelines, recognizing the restraints and constraints that exist, and providing for control, feedback, penalties for failure, and rewards for success. In other words, planning is a design for an entire system, and it requires regular evaluation of performance and continual adjustments to stay on course. It is a directive management process, not a permissive one.

As Peter Drucker says, it is easier to define what long-range planning is *not* than what it is. It is definitely *not* forecasting, that is, masterminding the future. It does *not* deal with future decisions; it deals with the futurity of present decisions. Finally, long-range planning is *not* an attempt to eliminate risk, or even to minimize it. "It is futile to try to eliminate risk; it is questionable to try to minimize; however, it is essential that the risks taken be the right risks." [2]

Long-range planning is a logistician's paradise. While an institution normally acts as a closed system, it is really a wide open one aspiring to be closed. In such a system, planning is the philosophical underpinning of the management process. Someone once called it an effort to develop a divine plan in harmony. A more earthly description is the traffic analogy: planning's purpose is to avoid collision. Plans are not made to be followed, however; they are made to be changed in order to operate effectively in an open system. They are comparable to semifinished goods in a manufacturing enterprise.

Long-range planning in an open system does not lead to anything definitive. There is a great deal of folklore in planning, and planners have been erroneously called "engineers of the future." But planning is only a part of the dynamic directive management process and cannot be absolutely blueprinted. An enterprise actually does not have a choice of whether or not to plan, for planning is an endless reconstruction and is necessary for survival. The "compulsively toilet-trained planner" tries to seal off and close the system, whereas the entrepreneurial planner tries to open it up and let innovative change in. The management process of an enterprise must include such an entrepreneurial dimension as a stimulus to growth and therefore must be vigilant to control the contraceptive forces which restrain innovative changes. The basic problem is to get management to feel comfortable in an open-management system—an area in which considerable research has been done by Alex Bavelas in his group studies.[3]

Planning is an activity in which a great number of people should participate. The management process is actually one of learning continuously while trying to improve results—that is, moving the enterprise further into the teeth of risk while reducing uncertainties. Karl Deutsch uses the river analogy as a model of the thinking that goes into planning. However, this analogy ignores a basic element, for the river does not learn. Planning should include a learning loop that is fed, servomechanism-style, into the process.

In planning, management makes preparations for changes in the environment, not only in operations but even in management philosophy. Planning's purpose with respect to innovations is to reduce situations of uncertainty to risk situations which can be probabilistically assessed for future operations. To improve planning, management must improve forecasting, both technological and environmental. There are basically three types of forecasts:

First, the statistical extrapolation of historical data.

Second, the subjective adjustment of the statistical forecast based on understanding of the future environment. When it is possible to reproduce the system in the form of mathematical models, it is possible to project alternate courses of events over time.

Third, the development of a venture simulation model. This involves the use of industry decision rules and associated time delays in combination with current industry data and management knowledge of future trends. It is perhaps a more realistic approach to forecasting and is often used in development situations. But such sophistication is not always possible, for basic data may be lacking.

The traditional image of long-range planners is that of a staff group gazing into their crystals behind locked doors. Everybody files a copy of his subplan, which is synthesized into a master plan by the central staff. It is then followed rigorously like a blueprint, backed by the full authority of top management.

A more modern enterprise mixes planners into every key-decision process. Planning, in the form of exploration of strategic options, contingencies, constraints, and probabilities, becomes a vital part of management. This system's approach allows for continuous reconstruction of plans within a loose general strategy in response to the flux of events.

It is particularly appropriate for dealing with development areas, where there are more uncertainties.

Why Plan?

People who count their chickens before they are hatched act very wisely because chickens run about so absurdly that it is impossible to count them accurately. *Oscar Wilde*

We can learn something from the various ways in which established companies have made major growth moves. Some have been content to enter new markets without much planning; others have utilized the standard approach; and still others have devised a more sophisticated strategy.

The company taking an unplanned approach moves into a given market at any time, with expectations based almost entirely on chance. While unplanned approaches have sometimes paid off, they are less and less successful in the modern business environment. They tend to waste effort, and fatal drawbacks often appear in the absence of any analysis of alternatives.

As a company realizes that the unplanned approach is unprofitable, it becomes more systematic. The classical procedure is to make a market-by-market analysis which involves home-office desk research followed by on-the-spot investigation by a marketing man. If his report is favorable, a task force field study is conducted before the final investment, postponement, or rejection decision. But even though this method may be textbook-perfect, it can sometimes yield critically misleading results. The reason is that no market is self-sufficient enough to remain independent of trends in world trade, technology, and opinion. The optimum decision for a single market might be disastrous because of events in other areas. These factors can affect nonprofit ventures as well as profit-seeking ones.

To minimize the risks of moving abroad, some of the more

sophisticated profit-oriented companies are creating new planning approaches. These usually begin with two fundamentals of good planning: analysis of the ultimate *objectives* of the enterprise, and the formulation of *alternate strategic plans* for the use of limited resources on a priority basis. In global planning, the ultimate objective for an enterprise making its first move overseas is usually to expand this foothold later without encountering numerous obstacles that were created by its first move. In this form of planning the chief executive must be willing to extend the scope of his responsibility. Before making the move, he should ask himself, "Where in the world should I invest this enterprise's time, manpower, and funds for the best long-term interests of our shareowners, customers, employees, and suppliers?"

In formulating strategy for innovative development, one of the modern corporation's problems is how to face its own organizational complexity, plan for it, and coordinate it. In the simplest form, the African village organization needed only one manager: the rainmaker. When no rain came, the rainmaker was buried alive and a new one installed. In modern times things are not so simple. As Casey Stengel once said at the close of a pennant-winning season, "I couldn't of done it without the players." And so it is with the job of innovative development planning—it cannot be done unless all individuals are oriented toward the team's objectives. The functions of the planning process for an enterprise seeking to introduce innovations might be summarized in the following manner:

First: Planning should strive to reduce uncertainties to risk elements.

Second: Planning should be an effort to eliminate surprise by competition, environmental changes, or external developments.

Third: Planning should provide a method for discriminating among multiple choices for investment of effort, time, and resources. Some overall strategy is necessary to conserve resources and to intelligently coordinate the company's projects.

Fourth: Planning seeks out and defines new growth or need potentials in market areas. The problem of the planner is then to evaluate the various options of conducting the business by extension, innovation, acquisition, and so on. He must also facilitate integration of the innovative activities with perpetuation of the existing business and further development of markets.

In international business, there has been a dramatic switch in business philosophy toward an overt internationalizing of the headquarters corporate staff and the assignment of worldwide product responsibility to home divisions or to regional areas. Managers have found that there is no single proper way for a profit-oriented international company to move into an underdeveloped area. They are coming to the conclusion that this "misery of choice" requires each manager to fit his own style, organizational architecture, and corporate commitment to the tasks ahead. The basic limitation is, of course, management experience and talent. Strategic planning for normative and innovative growth requires intelligent coordination of all efforts to help conserve vital resources.

Social Innovation

Sociologists have defined public-interest decision making as a "complex process of multi-lateral brokerage amongst legislators and public executives in and out of office." [4] Order in a free society requires that diversity and partisan differences be accepted as culturally legitimate. Social innovators must also cope with a persistent shortage of resources — natural and native — when planning for satisfactory social, cultural, economic, or political progress. Further, when the planning vehicle — the entire management organization — is involved in a social innovation, certain unique parameters overlie the otherwise challenging management process of planning within uncertain, changing, conflict- and tension-ridden business areas. These parameters usually include —

- An overriding purpose and objective of building society, as distinct from building an economic system.
- A determinative management faculty having prime motivations in

  (a) the form of altruism stemming from personal compassion.
  (b) service as a duty *quid pro quo* for the privilege of citizenship.
  (c) compulsive service related to critical national needs.
  (d) a yearning for participation in meaningful service activities – expressed as the "nobility of the service" ethic.

- An administrative faculty of management imbued with missionary zeal, scientific curiosity, enlightened self-interest, and/or outrage coupled with conviction, to quote Virginia A. Carollo of the staff of M.I.N.D.[5]
- Cultural gaps between varying value systems.
- A permissive style of management, or at best a coalition style; seldom a directive one.

These five parameters contrast with man's otherwise well-demonstrated motives: to "invade for gain," to be concerned for his personal safety, and to pursue glory, honor, and reputation.

Granted the special attributes of a socially oriented innovative effort, however, there seems to be no realistic justification for failure to employ good management philosophy and practice in its administrative functions. For example, technology transfer to a less developed country requires adaptation to that particular culture, and the "have-not know-how" adaptation process demands intelligent administration. The Peace Corps has what might be called a theological approach, in that certain "given" social beliefs govern its planning, coordination, and execution efforts. In a more directive approach, on the other hand, the determinative body (the board of trustees or directors) first decides what it wants to do and then marshals the administrative or directive management to examine all the realistic strategic options available to accomplish these objectives. Such an approach eliminates the belief in "what needs to be done" and channels effort into "what can be done." In the directive profit-motivated world, this is sometimes called a strategic market-back approach. It is suggested that the same approach has application in social innovation.

Strategic Planning

Without strategic planning, any effort becomes random, and decisions tend to be meaningless ad hoc choices. Strategic planning is a necessity because of competition, uncertainty, and change; it focuses on the organization's objectives in the light of these. Strategic planning is the economic use of resources, while tactical planning is the means for control. Further conservation of resources can obviously be achieved by intelligent collaboration with other social development efforts. Suggested steps in strategic planning are

1. Establishment of purposes and objectives.
2. Determination of planning premises: What kind of uncertainties, opportunities, and environmental constraints and restraints will there be? What resources are available? What is the timing framework?
3. The search for and examination of alternate strategic courses to achieve objectives.
4. Evaluation of these alternatives.
5. Selection of the best course.
6. Formulation of necessary tactical derivative plans.
7. Continuous assessment of competitive efforts and conflict pressures, with feedback for control and realignment of effort to cope with changes.

Given these seven steps of strategic planning, results will be enhanced if a marketing-oriented philosophy is adopted for approaching social-problem areas.

Planning must be accepted by the entire organization as only a part of a total management process which—

1. Requires a simple, fast, rational system to pull together the total resources of the organization into a strategy that maximizes effectiveness of the service which it is offering.

2. Needs to master the art of "marketing" its development and innovative efforts, which are offered in a changing situation of social conflict, tension, and differing value systems. This art includes skills for marketing to and around any constraining political structures.

3. Provides some type of coordination and incentive measurement to overcome the conflicts of interest which limit joint contributions among social and volunteer efforts.

4. Devises a tailored program to intelligently promote effective efforts for each innovative venture.

The basic conflicts encountered are predominantly behavior problems among people. Sometimes it is helpful to separate people into two groups: the masses and the few. The masses by and large resist change, but under pressure even they change with time. They are worried most about how to cope with economic depression and how to achieve basic security, and they have long looked for leadership to solve these problems.

The few, who are the leaders, can be subdivided into two groups: those dedicated to preserving freedom of choice for the people, and those dedicated to denying the people's freedom of choice. The marketing approach—whether it is adopted by a business, a voluntary organization, or a public service—is one that belongs to the first subgroup. It aims to preserve freedom of choice, because marketing leaders want to sell their ideas to the people and thus realize achievement. In contrast, the objective of the second subgroup of leaders is to exploit the masses by eliminating freedom of choice and substituting a philosophy of welfare and collectivism.

The innovating management needs to master the art of

marketing to the extremely important few in the socially sensitive areas. This requires careful coordination of the efforts directed toward these leaders.

With these concerns in mind, it is suggested that development and planning of an optimum strategy for innovation can be effective with socially sensitive organizations. A marketing-oriented task force operating under a strategic overall plan should pull all forces together to serve the total needs of an area. Such a project requires cultural and regional sensitivity, decisions that take account of the environmental implications, and a constant forward-looking operation on a planned basis. An evaluation, control, and coordination mechanism is essential to close the loop in the overall management process.

Technological Forecasting

In an article called "The Hazards of Prophecy," Arthur C. Clarke defines the two types of prophetic failures as failure of nerve (knowing that it could be done but believing it would not) and failure of imagination (not realizing it could be done).[6] Clarke contends that the most famous failures of nerve have occurred in the fields of aeronautics and astronautics, which are often considered the epitome of aggressive technology and modern management.

The year 1953 was a particularly disastrous one for those who indulged in aeronautical predictions. It was the fiftieth anniversary of the Wright brothers' flight, and publications produced memorial issues rich in articles reflecting on the past and looking fifty years ahead. A review and prognosis made by top aircraft and missile experts in the Aeronautical Science Institute's *Aero Engineering Review* expressed only one cautious hope for supersonic operational flight. Missiles were projected primarily as winged atmospheric vehicles, and there was no suggestion of massive ICBM threats or prediction that missile systems would take over many of the roles of aircraft. There was no reference to a coming space age (four years later, Sputnik I was launched). Not one of the 1953 article writers foresaw that within a decade the Institute for Aeronautical Science and its *Aero Engineering Review* would be out of existence. By 1963 the upstart American Rocket Society had become so large that the two were combined into the American Institute of Aeronautics and Astronautics, and the *Aero Engineering Review* became *Astronautics and Aeronautics.*

The word *astronautics,* virtually prohibited in polite engineering society before 1957, is now a household term. Although the space age burst on the startled world in 1957, almost all contemporary opinion strongly questioned the wisdom of placing man in space. Yet within four years a massive man-to-the-moon program was under way, and in eight years more it was completed.

Other notorious forecasting failures regarding computer technology, nuclear power, and chemical therapy could be cited. Even so, as the end of the century approaches we are inundated with year 2000 and year 2001 predictions. Another loophole the technical predictors always leave for themselves is the possibility of a technological breakthrough which will outstrip all forecasts. Clarke, however, suspects that "what's commonly become known as technical breakthroughs are quite predictable and failures to anticipate aggressive technical improvements are correctly recognized as merely failures of nerve." Many observers have noted that technological advances seem to occur at a frequency somewhat greater than one per decade. One of the explanations offered for this phenomenon is the concept of technological pressure — the idea that technology actually advances at an increasing, accelerated rate. Whereas our tendency is to be optimistic for the immediate future, we tend to be pessimistic for the more distant future. This is the natural slope of the "give a person a little responsibility and he turns conservative" curve.

With respect to technological pressure, the psychological effects can be imagined at the start of an advance. A proper technological decision is made, and the resulting program represents the thoughts of a good many innovators who are in tune with the technology of the day. It is almost certain, however, that during the succeeding year there will be attacks on the program. These will come from reactionaries wanting funds for totally different enterprises, as well as from other innovators trying to prove that they could do the particular job better. In other words, the "outs" will attack the "ins." At first the attacks will be ineffective, for reactionaries rarely stop a needed program, and technology will not have progressed sufficiently for other innovators to generate superior alternative programs. "After several cycles, however, two things happen: (1) reactionary attacks always continue but those innovators who honestly think they have a case become discouraged and go off to other fields; (2) the mental framework of the *ins*, perhaps originally quite flexible, has now solidified after they have been forced to defend the program many times. They have committed their reputations and begin to identify with the institution curve."

In what Clarke calls our "massive technological jungle," there is usually enough independent effort to guarantee that some change in the technological art will take place after the first few years. At that time the predators are likely to be different innovating outs from those who were unsuccessful a few years earlier. The new outs will be successful because they really have something to offer in the form of new technological perspective. The ins are not likely to realize that times have actually changed until it is too late. They may not even recognize the attack as a new and valid innovation but may dismiss it as just another reactionary threat. This is the concept of technological pressure according to Clarke, and it creates turmoil, revolution — and progress.

The continual increase in the curve of technological progress is easily misunderstood. There is a tendency to assume that progress is the exclusive result of scientific discoveries, but often the curve simply represents the growing application of available technology. In fact, some of the most significant developments have been the result of cleverly (or serendipitously) putting together readily available technology into new systems. As Clarke says, if the technological progress curve is used as a "pyramiding series of divine inspirations, there is a tendency to ignore its existence until such breakthroughs are publicized." If, on the other hand, the curve is viewed as the natural accretion of interacting knowledge from different fields, its behavior is more credible. Then again, some commentators tend to become nervous when a revolution is not in sight. They develop increasing worries about the surprises that are being prepared elsewhere in economic and military breakthroughs.

Ideally, of course, system innovators would base new systems and programs not on the state of the art available when planning is started, but rather on the art available at the operational date. Clarke says: "By taking a calculated risk as to the rate of advancement of the technology in the same way that a commercial entrepreneur takes calculated risks as to the development of the market place, the dealer in advanced technology earns his spurs."

Technological Innovation

The intricate process by which technological innovation takes place has been studied for over 20 years. A number of surveys, such as the study conducted by Erich Jantsch for OECD, have attempted to discover the basic conditions which favor technological innovation.[7] At least two of the surveys found a similar pattern. General Electric's Tempo Center for Advanced Studies, on the basis of 75 important technological innovations in the twentieth century, emphasized the following factors:

- The innovator had a clear purpose.
- Financial resources were available.
- An information base existed.
- The innovator was forced to learn a great deal — a factor which seems to favor outsiders who must learn about the field.

The survey conducted by Arthur D. Little which evaluated 63 successful research and development "events" in the history of six complex weapons systems found that these systems depended on many small inventions; only two major inventions contributed to their development.[8] It also concluded that an "adaptive" environment in contrast to an "authoritative" one is probably a prerequisite for successful development. The following factors were seen as most important:

- A clearly formulated need.
- Availability of resources to be committed *at once* (a delay of even one or two months has a discouraging effect—people stop producing ideas—and a six-month delay proves to be a terrible burden).
- An experienced body of people.

A study of the nineteenth-century Industrial Revolution supports the observation that the basic laws governing innovation have changed little over time.[9] The revolution did not take place in France, where the science originated, but in Great Britain, where more favorable conditions existed. These included the entrepreneurial spirit, a mobile labor supply, good communications, a liberal political system, and a market. The socioeconomic context of technological innovation is now apparent in the current automated industrial revolution. Europe has recently recognized the importance of this context, and significant innovative trends are taking place there.

These evaluations imply strong normative as distinct from exploratory thinking, and there has been misunderstanding about this, reflected in the phrase, "America has finders, not seekers," used by Erich Jantsch in his OECD study.

The problem of limited resources has arisen during the past quarter century, during which the opportunities for innovation have outstripped financial and manpower resources and the "misery of choice" has become a problem. According to a McGraw-Hill survey in 1965 and 1966, two-thirds of American industry as a whole considered financial and manpower constraints as the major obstacles to more research and development.[10] McGraw-Hill judged that the average rate of innovation in U.S. industry (measured by the sales value of products introduced each year which are either new in nature or new to the market) ranges from one to ten percent.

Technological innovation is more than ever due primarily to clear normative thinking—and any technological forecasting will include a strong normative as well as an exploratory component. This is also true of technologies that were "dormant" for some time in the absence of a clear incentive.

A 1969 study prepared by the Illinois Institute of Technology traced the scientific and technological background of five innovations which have had a major impact on the technological world.[11] This survey was somewhat of a counterweight to the Defense Department's controversial Project Hindsight report, which emphasized the links between basic science and productive technology. According to the Illinois study—

- Twenty to thirty years elapse between basic scientific discoveries and their commercialization. An innovation is conceived nine years before its realization.
- The existence of a diverse fund of knowledge is essential to the development of an innovation.
- Mission-oriented research is a dominant factor in development of products such as the equipment needed by the Defense Department.
- Only about 10 percent of the key events in an innovation occur during development after the original idea.
- Only about 20 percent of the key events are mission-oriented.
- Over 70 percent of commercial successes spring from innovations originated in nonmission research.
- About 90 percent of nonmission research is accomplished before conception of the product—that is, without insight into conceptions or innovations to which it will ultimately contribute.

The technological developments studied in the Illinois survey were the pill, video tape recorders, the electron microscope, ferrites, and matrix isolation. In another field, that of plastics, the pattern of innovation seems to bear out the Illinois conclusions. Although many plastics were already known in the laboratory in the 1920s, their commercial development came only after the rise of petrochemistry. Polyethylene and polyester chemistry are good examples. Isocyanates, invented in 1937, became an innovation only in 1962 with the development of appropriate molding machinery.

Another factor of utmost importance in innovation is the structural changes in industrial patterns caused by the invasion of one (usually stagnant) sector by another (dynamic) one. An ADL report describes the invasion of the textile industry by the chemical sector and the chemical sector by the petrochemical (and vice versa).[12] The industry whose position is independent with respect to the raw material has an added incentive here. Three factors have revolutionized the agricultural-chemical sector since 1963: invasion by the oil companies, the trend to bigger farms in the United States, and the backward integration of the food industry, which influenced farming methods. The gradual replacement of natural materials by synthetics may be regarded as another form of invasion. The recognition that the human population grows faster than the animal population, and the prediction that people will increasingly shy from employment in unpleasant types of work, such as the preparation of skins, were at the bottom of Du Pont's decision to develop artificial leather (Corfam).

Economics of Innovation Planning

Companies in the United States obtain an identifiable return from only 45 percent of the money they spend on R&D, judging from the experience of firms consulted by the American Management Association.[13] For each $100,000 invested in research projects, some $55,000 does *not* lead to products or services contributing to the company's profitability. If it is assumed that the average company earns 10 percent on sales—usually an attractive rate of return—it must sell $550,000 worth of its products to recover the cost of unproductive R&D. This finding was uncovered for AMA by Dr. Burton V. Dean. There appears little doubt that improvement in the processes of generating, screening, evaluating, and selecting R&D projects is called for.

To provide the means for estimating the impact of R&D on corporate performance, and to supply needed information on proposed and current projects, extensive evaluation of completed and terminated projects is being performed by technologically based firms. Ten years ago, such evaluations were made by fewer than one-fourth of innovation-producing companies. Now it is estimated that more than three-quarters of them carry out these analyses. Postaudits of projects take account of costs, completion time, capital investment, results of the conformed cost savings, and so on.

The companies surveyed by AMA reported that, on the average, a project has a mean life of 1.6 years. Across companies, however, a standard deviation of this value was 0.8 years, indicating a wide variation. Again taking an average, the firms reported that 6 percent of their projects resulted in some significant technical achievement, 33 percent led to a process or product modification, 31 percent yielded a new process or product, and, as mentioned, 45 percent resulted in an improvement of the firm's profitability position.

Research is too often considered a corporate status symbol; management excuses R&D from compliance with the performance and economic standards imposed on other functions. From 1940 to 1965, the Stanford Research Institute's index of research costs rose 380 percent, while labor costs in general rose about 50 percent. A 1966 study sponsored by the Joint Economics Committee of Congress and conducted by R. G. Gilfillan further dramatizes rising R&D costs: After developing data to determine long-range R&D trends, the study found that inputs—people, material, and capital investment—had increased 340 times between 1880 and 1960.

Yet output—the results of R&D—had increased only 105 times. As William K. Hodson says, if expenditures for research in the United States continue to double every seven years, which they have done so far, over half our workers will be employed in some phase of research by the end of the century.[14]

One of the reasons these costs have grown so rapidly is that research is surrounded by folklore. People tend to think of it as an end in itself, needing little or no economic justification and no discipline. The research mystique is accepted so widely that one New York brokerage firm regularly recommends stocks on the basis of research dollars per share. Another misconception is that researchers perform best when they are uncontrolled. Still another, according to Hodson, is that people in research centers are mainly engaged in research. Management deludes itself by assuming that everyone in these think tanks is actually thinking eight to ten hours a day. A study of 100 scientists and engineers in the March-April 1958 *Harvard Business Review* found that they were performing at a composite efficiency level of 8 percent based on a 100 percent standard. More than half the activities of a research center consist of routine services, answering letters, handling complaints, and perfecting a process—in other words, doing nothing to increase profits, while increasing the cost of the operation.

In a 118-week study of a Swedish research center, Hodson reported that 30 percent of the time was devoted to "assuring the scientists that improvement rather than scrutiny was the objective." The 330 research workers were interviewed on a random basis, and more than one-quarter of those questioned were doing no work at all. At least 65 percent of the staff were involved in routine operations. Interviewers also found that only 10 percent of all R&D personnel were engaged in pure research. Since 65 percent of the center's staff were engineers and technicians and only 35 percent were clerical employees, the level of productivity was not impressive. Based on the results of this survey, a detailed plan of organized research was set up at the research center. Research costs as a percentage of sales dropped drastically, and the trend on research expense, which had been doubling every five or six years, reversed itself. There was no need to add new personnel for at lease five years, whereas the center had been adding 25 engineers a day! The corporate growth rate has since risen from 4 to 15 percent per year —much of this increase due to the revitalized, newly organized research program.

5
Environmental Aspects of Innovation

Science never solves a problem without creating ten more. *George Bernard Shaw*

WHILE innovation is generally regarded as a blessing and the fountainhead of science and the arts, it is also a curious phenomenon in that it makes human beings both slaves and masters. For wherever innovation occurs, giving people more power over their environment, change invariably results, forcing them to innovate in order to respond. The cycle is endless.

In the United States, the opposition to technological innovations which erupted in the 1960s and continues with increasing violence in the 1970s has often been aimed at specific targets: nuclear weapons testing, insecticides, supersonic transport, location of atomic power plants, and new superhighways. Beyond these, general opposition to the environmental effects of industrialization is found throughout the world. As a result of the serious problems arising in the United States, Congress in the past few years has produced landmark legislation on environmental pollution and consumer protection.

Lord Ritchie-Calder of Balmashannar, former professor of international relations at Edinburgh University, charges that the "great achievements of *homo sapiens* become the disaster-ridden blunders of Unthinking Man." [1] He relates a sobering story about the type of mistake which comes from the very best intentions to innovate. In the Indus Valley in West Pakistan, the population is increasing at the rate of ten mouths to be fed every five minutes. In the same five minutes in the same place, an acre of land is being lost through waterlogging and salinity. This valley is the largest irrigated region in the world, with 23 million acres artificially watered by canals. Founded by the British in the nineteenth century, the Indus irrigation system has been extended through externally financed programs supervised by hydrological experts. Over the years inadequate drainage has caused low areas to become waterlogged, the water table has risen, and leaching salt accumulation has led to a tragic poisoning of crops. President Ayub appealed to President Kennedy, who responded by sending a high-powered mission which determined that it would take 20 years and $2 billion to repair the damage—more than it cost to create the entire canal installation in the first place.

Concern in the United States about adequate consideration of the total consequences of technical innovation is sufficiently widespread that the National Academy of Science had studied the problem and summarized its findings in a report titled *Technology: Processes of Assessment and Choice*.[2] The study panel, which concentrated mainly on the process of making technological decisions at the federal level, observed that technological change in this country has been governed primarily by market mechanisms. Damage done to individuals, and adverse consequences to society as a whole, have been largely ignored.

The Academy panel searched for objective means to establish what it called "deleterious secondary consequences," or what economists term external diseconomies. The findings are not very encouraging. The panel backed an approach to evaluation based on "a network of mechanisms," since no master plan for technological assessment was thought desirable or possible. This constellation of organizations would create a focus and a forum for responsible technology-assessment activities throughout the government and private sectors. Mechanisms, the study group suggested, should be subject to independent, external criticism and should be given the structural resilience to change with experience. The panel was adamant that the assessment organization should be insulated from policy-making power — that its responsibility should be to preserve neutrality and to study and recommend but not act.

The NAS report is rational and intelligent. In essence, its prescription for dealing with innovations is an extension of the postwar pattern which brought university scientists into working contact with government as advisors, as researchers, and sometimes as upper-level civil servants. Although this alliance in the past added new dimensions to the federal bureaucracy and proved generally satisfactory to both scientists and government, the record of government in guarding the public against the negative effects of technology has not been inspiring. During the last decade, according to the NAS report, information about the impact of technology on society has generally been made available through "the efforts of superior provocateurs like Rachel Carson and Ralph Nader, through actions of indignant individuals or groups often campaigning in the tradition of Don Quixote." However, the panel does not seem to give much weight to these activities as an innovative method of accomplishing an important mission. More recent pressures by such nongovernment groups with their "quixotic" acts of defiance seem to be more successful. It is significant that President Nixon's 1970 budget message set forth an increased amount of federal funds to cope with the problems caused by adverse effects of technology.

It is generally agreed that science and technology have introduced so many innovations in recent years that massive social changes must take place in order to accommodate them. While these innovations present many problems, they also offer opportunities for further innovation that may leave us better off in the long run. Some of the problems have their roots in such fundamental issues as population growth, which in itself is a result of innovative procedures in food, medicine, and the like. The population explosion in turn has contributed to urban problems, transportation problems, pollution problems, and so on. The technological advances did not create the inequities and conflicts so much as they made them immense, intractable, and unendurable.

The human race has so far survived by its ability to innovate, and it should be able to continue surviving. The key phrase, however, is *so far.* In the modern age, traditional social and economic growth styles can no longer adapt to the enormous changes caused by such forces as nuclear energy or population expansion. If we are to innovate in the social and intellectual spheres in order to correct and respond to these massive technological innovations, we must first learn more about the innovation process itself and the characteristics of the innovative person.

It is worth mentioning that perhaps the only real source of power in the world is the gap between what is and what might be. Adiabatic change, to borrow a physics term, involves change without loss of heat—that is, a peaceful transformation. Such is needed to accomplish the management of social changes, if this is possible.

The Changing Environment: Three Trends

At least three major developments, primarily technological, have had a dramatic impact on our environment in recent history. They are the expansion of knowledge, the advances in science and technology, and the trend toward urbanization. The dislocation of our tangible and intangible resources, and the problems of conserving them and improving our standards of living and our progress in industrialization, presents overwhelming challenges. But solution of the problems first requires their definition.

Knowledge Expansion

There may be no new things under the sun, but the existing forms are certainly undergoing successive mutation at a rapid pace. These changes have burst full-blown upon the younger generation with an impact that is not adequately perceived. Barnaby C. Keeney, former president of Brown University, says that while the generation gap has been with us since the days of the prodigal son, "it is no longer so easy to grease it over with calf fat, because the gap is not simply between the ages of man but ages of society."[3] Today's changes are notable not only for their impact but also for their rate of acceleration. Taking the speed of transportation as an example, Keeney points out:

> For about 396,000 years man could move as fast as he could walk or run, then for 3,999 years more, as fast as a horse could carry him or pull him in a cart, or a wind-driven ship could take him across the sea. By the end of the 19th century, the railroad could move him at 60 mph, an increase of 56 miles per hour in 400,000 years. By World War II, 40 years later, the speed had increased by five times to 300 mph. Now he can travel 18,000 miles an hour more. Almost all this change has come in the last 30 years. The effect on our concept of time, our concept of society, our concept of the world in which we live has been enormously important.

The pressures from exponential increases in knowledge are such, educators say, that there is about 100 times as much to know now as there was in 1900. Knowledge has been defined as information acquired by some people for the sake of knowing it, and by others for the sake of telling it. By the year 2000 there will be more than 1,000 times more knowledge of all kinds than there is now; the problem will be not so much to know it or tell it but to record, classify, store, search out, teach, and, let us hope, use it with some discrimination and effectiveness.

It has been claimed that what the railroad did for the second half of the nineteenth century and the automobile did for the first half of the twentieth, the knowledge industry will do for the closing years of this century in the United States. This augurs a shift from a society based on natural resources to one based on human resources, of which the most important will be ideas. *Fortune* has described the U.S. $195 billion knowledge industry as the biggest growth industry of all, one that is growing twice as fast as the economy itself.[4] The production of measurable knowledge goes on expanding faster than the economy because it generates as well as feeds on economic growth. The growth of the knowledge industry is a major factor in rising productivity in the United States and, although perhaps on a different time scale, in other areas of the world.

Scientific and Technological Advances

That's the biggest fool thing we ever have done . . . the bomb will never go off and I speak as an expert in explosives. *Admiral William Leahy* in a letter to President Truman

Advances in recent years have followed one another so rapidly that people have learned to accept them equally fast. World reaction to the second lunar landing, for example, was almost blasé; men on the moon had become somewhat commonplace. It took the cliff-hanger of Apollo 13's recovery flight to refocus public attention on this technological feat.

The pace of innovation is also shown by the fact that in the late 1960s, 80 percent of U.S. college graduates entered positions which did not even exist before they were born. Furthermore, the country is now producing over 2½ times as many goods as it did in 1929, but using only 20 percent more manpower to do it.[5]

The term technological advance, or technological change, usually means one of two things: First, the introduction of new arrangements in the production or distribution process which make possible new or improved products or services. (The basic characteristic of technological change is that it permits resources to be utilized more efficiently. For a given amount of output, less capital, labor, and material inputs may be required; or else the same amount of resources may allow greater output to be produced.) Or second, it may mean the introduction of new equipment, processes, or products. (This can also be considered as the culminating step in a sequence that extends over a long period. The accumulation of knowledge underlying such a technological advance may represent the work of many scientists, inventors, and engineers.)

Expected savings from a technological change compared with the cost of existing technology must be large enough to induce the user to make the investment. Once the profitability of the change is proved, acceptance by most of the industry tends to accelerate — and by definition an innovation takes place. Adoption of a technological advance in an industry, however, depends on many nontechnical factors. Investment decisions by individual firms are influenced by their commercial outlook, capital requirements, competition, management philosophy, attitudes toward the unions and the public, government regulations, economic situation, and the like. As outlined in a recent U.S. government study, the pace of technological change is closely related to the rate of gross investment, the level of economic activity, and the changing structure of production and demand.[6]

The rate of technological advance is conventionally measured in terms of productivity, usually output per man-hour. However, while technological change is an important factor in productivity gains, it is not the sole factor. Productivity is also influenced by nontechnical developments, such as capacity utilization and long-term improvements in management and in the skill and educational level of the workforce. An innovation may reduce costs so much that sales increase more sharply than the reduction in labor requirements. Or the innovation may permit work to be done that hitherto was uneconomical, in which case additional employees may be hired rather than existing ones fired.

Technological advances have been taking place for a long time, certainly as long as there has been any effective notion of progress. George Bernard Shaw believed that "progress is impossible without change; and those who cannot change their minds cannot change anything." Since the eighteenth century there have been at least three great changes of mind effected by technological advances.

1. The early age of mechanization, which began in the 1700s with the introduction of power-driven machinery for spinning yarn. This change displaced the handicraft worker and gave rise to the concepts of the factory system.

2. The age of mass production, when the factory worker's job became faster as conveyors carried standardized products and power production was transferred from steam-driven shaft and belt systems to central electric-power-generating stations. The mass production concept became one of the central facts in our way of life.

3. The age of science and technology, which burst on the United States after World War II with massive support of research and development by the federal government and by private corporations. The technological fallout and its effects on society are forcing major social innovations.

American industry now supports this technological continuum by spending $7.50 on research and development for every $100 of civilian manufacturing output, while the Defense Department spends $54 for every $100 of military procurement.[7] R&D already has brought such unforeseen developments as electronic computers, nuclear energy, jet propulsion, space technology, and automation of industrial processes. The continued rapid growth of population and markets, increasing foreign competition, and pressure for higher living standards are stimulating even greater efforts to improve technology.

However, with the public outcry over the bad effects of this technological continuum, the future probably will be characterized by an expanded emphasis on social innovations to cope with these effects. Such innovations will include improvements in the education and training of young people, measures for retraining and increasing the mobility of the management and labor forces, and more adequate provisions maintaining the income of the unemployed. Social invention will also involve labor and management in the creation of new collective bargaining approaches to deal with problems of technological change. Behavioral research will be an important factor in social changes made in the future. Innovators will create new concepts of organizational design, and new approaches to the interface that such systems have with all environments. In their turn, these and other social changes will constitute a new framework for future technological and social advances.

According to a study of technological trends in major American industries, nine are identifiable: [8]

1. Computerization of data processing.
2. Greater instrumentation and process control.
3. Trend toward increased mechanization.
4. Progress in communication.
5. Advances in metalworking operations.
6. Development of energy and power.
7. Advances in transportation.
8. New materials, products, and processes.
9. Managerial and related techniques.

The 40 industries covered in this report employed 33.8 million people in 1964, that is, 58 percent of the nonfarm working population.

What about the technology of the future? Herman Kahn and Anthony J. Wiener, in their framework for speculation on the next 33 years and beyond, have analyzed changes in continuity in the perspective of history.[9]

Relatively *near-term* developments (predicted for the last third of this century) include improved chemical control of mental illness, capability to choose the sex of unborn children, and interplanetary travel.

Not so near-term innovations (which are less likely to occur by the year 2000 but might appear within that century) include the technological equivalent of telepathy, the conversion of mammals to fluid breathing, and the artificial growth of organs.

Less credible but still not impossible, according to the authors, are innovations that could occur after the dawn of the twenty-first century: substantial lunar or planetary colonies, antigravity, routine and practical use of extrasensory perception, very low-cost electric power, and many others.

When practical managers contemplate such possible innovations, it is prudent to remember that what is done now will profoundly affect what lies ahead and that it is a great challenge to be ready to manage these potential changes, particularly in the social area. Too, the advances from the technological side must be managed so that they will not be unduly curtailed or repressed because of the disadvantages and perturbations they introduce into our social system.

Thus the management of new technology is and will remain a critical requirement as technological innovations burst upon us from many sources at an accelerating pace. The business leaders of the future must be able to manage mind-changing new technologies within an increasingly sensitive environment. It will be a perpetual learning process, and the sooner managers understand the dynamic nature of the situation and the anatomy of innovation itself, the better they will be able to deal with changes and their consequences.

Urbanization

The third major trend in the age of science and technology is the clustering of people far away from the sites of natural resources and in unnatural conditions of "captivity" within the teeming environment of our crowded cities. The social hazards of this vast impersonal community are changing the life style of people who seek the exhilarating opportunities of urban existence but are forced to deal with its dangers and disadvantages.

Migration to metropolitan areas has been going on for a long time, but today we are becoming urbanized at increasing speed. Some of the causes have their roots in scientific and technological advances: the ability to produce more food on the farms with fewer people, the improvements in communications and transportation, our increased leisure, the generally affluent society in certain world areas, and the population growth.

Professor Kingsley Davis of the University of California at Berkeley has shown that 38 percent of the world's population are already living in what he defines as urban places.[10] Over one-fifth of all people live in cities of 100,000 or more, while about one-tenth (in 1968, 375 million people) are in cities of a million or more. He estimates that it would take only 16 years for half the people of the world to become city dwellers, and if the trend continues, only 55 years for all the people to congregate in cities. Professor Davis predicts that the present population will increase five times within the life span of a child born today. This population of 15 billion could all be living in cities of more than a million people, with the biggest city composed of 1.3 billion inhabitants. Such a city would have 186 times the population of Greater London.

Growth and Innovation

In the United States

In a report on technological innovation, the Department of Commerce has taken a look at the histories of three U.S. industries — television, jet travel, and digital computers — that were nonexistent in 1945 but have contributed significantly to the nation's growth over the past 20 years.[11] In 1965 these industries were adding $13 billion to our GNP and were employing 900,000 people. The report examines specific companies within them that committed themselves to innovation and have a distinguished record of introducing products, processes, and services as a way of business life. The histories of Polaroid, 3M, IBM, Xerox, and Texas Instruments show that while the average annual growth of the GNP over the period 1945–1965 was 2.5 percent, the average annual net sales growth of these companies was 13 to 29 percent, making an average for the group of nearly 17 percent. (Texas Instruments was excluded despite the fact that it had the highest growth rate of all, since data were not available for 1945.) At the same time, the average yearly growth in jobs ranged from 7.5 to 17.8 percent.

Another significant effect of technological innovation is the improved position of a country in international trade. This, of course, is one of the major reasons why the federal government is concerned to promote invention and innovation. The technological balance of payments, an important element of our international balance of payments, is an international account reflecting payments for technical know-how, patent royalties, and the like. A study of this account conducted by OECD showed that in 1961 the United States was receiving roughly ten times as much in technological payments as it was paying out to other nations.

Technological change also affects international trade in more subtle ways, as reported by the Department of Commerce. For example, displacement innovations can be distinguished from new innovations. A new innovation is one like the Xerox machine or the electronic computer — a product for which no substitute existed before. Displacement innovations are those which push out existing products or processes, as illustrated by the invasion of the cotton and wool fiber markets by synthetic fibers. The drop from $187 million to $125 million in U.S. exports of cotton and wool during the period 1956–1965 contrasts with the growth of synthetics exports from $158 million to $241 million.

In Europe

According to *Business International*, the major challenges facing the European Economic Community today include foreign investment, the technological gap, and the relation of these to the European environment for innovation.[12]

The role of foreign investment in EEC is under continuous scrutiny by national governments, EEC firms, unions, antitrust authorities, economists, journalists, and intellectual leaders on both sides of the Atlantic. EEC is also actively searching for a solution to the technological gap and the concomitant managerial gap.[13] Europe's economic recovery, which began in the early 1950s, encouraged U.S. firms in increasing numbers to set up plants and marketing operations there. In October 1969, however, the Commission opposed a regrouping of the heavy electrical equipment industry in EEC when Westinghouse attempted a series of takeovers. For some time, U.S. investments in EEC had been directed toward growth sectors, and it was these investments that spawned the adverse reaction from the Commission, which then began a program of aiding EEC firms through subsidies and other policy means. Areas currently receiving this attention are electronics, paper and pulp, shipbuilding, steel, and textiles. Another sector is the petroleum industry, where financial support for oil prospecting is limited to "community enterprises," that is, EEC companies.

American participation in EEC markets ranges from approximately 10 percent in a few chemical commodities to about 80 percent in the computer field. It varies widely by country but has been increasing at an impressive rate as the United States puts a greater percentage of its overseas direct investment into Europe. In the past 16 years this investment has doubled and now represents almost 30 percent of the total world overseas direct investment by American firms.

Until 1960, U.S. direct investments were a minor part of gross capital formation in EEC. According to the *Business International* report, in 1966 our capital formation accounted for about 9 percent of total EEC capital expenditures in the industrial sector and about 12 percent of total EEC fixed asset outlays in mining and manufacturing.

Another challenge facing the European Economic Community concerns the technological gap, particularly in growth industries. The United States has invested most in chemicals, electric and electronic equipment, and the pharmaceutical industry, all of which account for 68 percent of our total direct investment in EEC from 1958 to 1965. According to OECD reports, some aspects of the gap are that 17 percent of all patents registered in Western Europe during 1958–1963 were of American origin, and in 1963 U.S. net earnings from foreign royalties and technical assistance fees totaled $754 million.

In absolute terms, the United States spends about six times as much as EEC on R&D, and four times as much when R&D is measured on a per capita income basis.

Small-Business Aspects

The U.S. business environment is characterized by the continual emergence of small, technically oriented or marketing-oriented firms started by entrepreneurs. These men have been encouraged to satisfy their desires for innovation by the availability of venture capital through a new breed of financial organization, by some favorable tax regulations, and by the tremendous (200 million people) and relatively uniform American marketplace. (Europe's consumers, although they are slightly more numerous, do not yet represent a uniform market.)

This country's management style, relative freedom of action for the businessman, and minimum obstructions in the way of tariffs, language, and legal and currency problems provide a fertile field for the entrepreneur and innovator. Certain progressive large firms here are also becoming actively concerned at boardroom level with the unique management challenge of nurturing innovative projects, whether these are offered from outside or are generated inside but frustrated by corporate inertia.

The relatively closed system and inexorable momentum of the conventional large, successful company make innovative projects hard to manage unless they are "planted" outside the normal corporate preserve, or unless corporate policies are significantly modified. The dead hand of central control and the corporate overhead burden can quickly swamp an uncertain new project. A large-company "risk avoidance" management style cannot be applied to a budding enterprise faced with uncertainties of all types. Nor can laborious mechanisms for obtaining policy decisions and financial resources be adapted to risk situations. Consequently, outside financial firms have sprung up in the United States to help the inventor, the innovator, and the entrepreneur — men who need funds and freedom to fail or, they hope, to create a business and personal estate of their own by total commitment to some ingenious business project. This venture capital movement, covered in more detail in Chapter 9, is a significant factor in the environment for innovation.

As it moves to control industrial activities which adversely affect the environment or the well-being of society, the federal government has recognized that small business needs help in complying with the new regulations. One bill (S-1649, introduced April 26, 1971), is designed to "amend the Small Business Act to authorize assistance to small business concerns in financing structural, operational, or other changes to meet standards required pursuant to certain Federal or State laws." Although such financial assistance may eventually be available to the small businessman, he also needs advice on how to modify his process to eliminate undesirable side effects without eliminating the profitability of the entire enterprise.

In the case of the Vermont cheese industry, for example, federal and state water pollution legislation forced the cheese manufacturers to discontinue dumping whey into rivers and streams as of 1969. Learning that these laws would result in bankruptcy or shutdowns for some cheese producers and curtailed operations for others, the program director of Vermont's Office of State Technical Services asked the University of Vermont to investigate the feasibility of converting the liquid whey into a dried, edible product.[14] The results of the study were favorable, and with these findings he set about to attract to the state a major food processor with the required technology to operate the drying plant, organize the 18 cheese manufacturers into a cooperative group to supply whey, and apply for $2 million in federal grants and loans.

The project, which was directed and coordinated throughout by the program director, had these economic effects:

1. 54 new direct jobs at the whey plant, which is estimated to have an annual payroll of $464,000 based on sales of $1,440,000 of dried food-grade whey per year.
2. 51 jobs saved at 4 cheese plants which would otherwise have closed (annual payroll of $255,000).
3. 28 new jobs created through the increase in cheese production (annual payroll of $395,000).
4. $454,000 in new federal, state, and local tax revenues annually as a result of the industrial activity.
5. 120 new jobs in secondary employment in Franklin County, an area with serious unemployment.

The Vermont whey plant was one case in a nine-state study of technology transfer to industry. It is significant that a small number of successful technology transfers to industry accounted for the largest economic effects, with field services being the dominant STS activity. In all nine states which were visited, it was found that successful instances were only a small percentage of the total number of companies where some form of service was rendered. Although further improvements in the operation of a technology transfer program can be expected to increase the number of such successes, the general distribution pattern is likely to remain the same; this is indicative of the high risk of innovative activities. The fact that the risk is particularly difficult for small business to bear justifies federal support of a program of technology transfer, which spreads the risks across a large number of companies in order to obtain a favorable ratio of return on investment. No other institution gives similar assistance to small business on a national scale, and on any smaller scale such institutions have not found it profitable to serve the needs of small business.

Enterprise Growth and Complexity

As an enterprise grows in size, its system, its dynamics, its constraints, and the relative importance of factors necessary for survival and success shift radically. The complexity caused by people and paperwork is a troublesome problem for the manager concerned with innovative growth. While Parkinson's third law postulates that expansion per se means complexity and complexity means decay, in statistical theory it is clear that this complexity can be characterized partly by the number of people contacts which require coordination. Contacts jump from about 100 possibilities for an executive with five people reporting to him to over 1,000 when he has eight subordinates. The point where decay begins is uncertain.

The amount of internally generated and circulated paperwork is another factor in organizational complexity. As the company increases in size, its efficiency generally first rises and then falls off in inverse proportion to the number of employees. The end is reached when everyone spends all his time reading everyone else's reports. Some German theoreticians claim that by not looking at their mail, 4,800 workers could turn out as much work as an infinite number of workers who do read their mail. The same scholars have determined that a purely administrative organization becomes self-sufficient as soon as its staff numbers 1,000. From then on, it generates sufficient internal correspondence to keep itself busy without any incoming mail or external contact of any kind.

The administrative intensity of an organization has been quantitatively defined by Louis R. Pondy as the number of managers, professionals, and clerical workers divided by the number of craftsmen, operatives, and laborers.[15] Using this definition, a mathematical model of a manufacturing firm is derived by treating administrative personnel as an input factor analogous to labor and capital. The model assumes that administrative intensity is set so as to maximize profit or, more generally, to maximize the dominant manager's utility function. It was tested in Pondy's study against data for a sample of 45 manufacturing industries in the United States. Consistent with the predictions of the model, administrative intensity was found to decrease with organization size and to increase with functional complexity and with the separation of ownership and management. Administrative intensity was also positively related to value added per production worker, and negatively related to the average salary of administrative personnel.

In the 45 manufacturing industries Pondy studied, the number of administrative personnel per 100 production workers varied from 8.7 in the logging industry to 131.1 for the drug industry, with a mean of 37.7 and a standard deviation of 28.8. The theory is that the number of administrative personnel employed in an organization is chosen in order to maximize the achievement of goals of the dominant management coalition. The relative size of the administrative components of an organization is treated as a variable subject to management discretion. This treatment is in contrast to the traditional approach, which assumes that administrative intensity can be technologically determined by the task of the organization or by other situational or structural characteristics. According to Pondy, it has been argued that administrative intensity is determined by (1) "task complexity and division of labor creating problems of coordination which, in turn, require more administrators, (2) spatial dispersion of organization and multiple departments making coordination more difficult and, in turn, necessitating more administrators than with spatially concentrated members, (3) variability and heterogeneity of the task environment, which requires large numbers of administrators to standardize, stabilize, and regulate input-output transactions, so that the core technology organization cooperates in an environment of technical rationality."

The first argument is fairly well supported by numerous studies. The second receives some support from data on tax-supported universities and public schools, provided size is controlled. However, certain studies of local labor unions contradict this hypothesis. The third argument concerning variability and heterogeneity is generally consistent with analyses of American corporations. Recent research suggests that administrative intensity is lower (in other words, there is a larger span of control and fewer hierarchical levels) in organizations or organizational subunits facing unpredictable environments and technology (for example, basic and applied research divisions) than in those facing relatively stable environments and technologies (such as sales and production divisions).

> With the separation of ownership and management, management's motivations may not be strictly oriented toward profit maximization. Management may be motivated to increase administrative personnel beyond the optimum profit point. After all, non-owner managers do not share directly in increased profits, although they may share more directly by spending those profits on subordinate administrative personnel to enhance their prestige, to make their jobs easier, and so on; that is, they value hierarchial expense, per se, as well as profitability. Or management may attach a *negative* "expense preference" to administrative personnel. Owner-managers, for example, may not wish to weaken control of the organization by bringing into the administrative structure persons outside of their families. They may even be willing to accept a lower profit as a result of being under-administered in exchange for maintaining close personal or family control of the organization. According to Pondy, the dominant manager's utility function is defined to include both profit and hierarchial expense.[16]

This study makes the point that while administrative intensity in organizations is a function of both economic and technological variables, it is also strongly influenced by managerial motivations and patterns of ownership.

6

Combating Entreprenertia: How to Organize for Innovation

In science the important thing is to modify and change one's ideas as science advances. *Claude Bernard*

THE great barrier to major entrepreneurial innovations within the modern corporate framework is the organizational hierarchy itself, which operates to protect the status quo and fend off attempts at innovative change. This hierarchical hangup is manifested by the inertia of the managers in power, whose tendency to neglect, sequester, or smother the entrepreneur might be termed "entreprenertia." The high resistance energy in the hierarchy to entrepreneurial endeavor comes from the intellectual, emotional, political, and economic investment of the individuals who have institutionalized their place in the corporate system. A formidable anti-innovation and anti-entrepreneurial ambience develops, and it takes a torporific toll of the entrepreneur who wants to

114

assume the risk in management of an innovative undertaking. He is blocked as a peace disturber, as a threat to the status quo, or merely as a member of the young boys' network.

New concepts and different styles of management that are worthy of consideration have always been available and always will be. The problem is that they are often held in abeyance, consciously or unconsciously, by the inertia of the establishment. When Francis Bacon warned, "He that will not apply new remedies must expect new evils; for time is the greatest innovator," the manager's job and the environment were relatively static. With the impact of technological innovations and the social consequences calling for new remedies, the inflexibility of the organization becomes even more of a liability.

Kainotophobia, the fear of change, and self-interest motives are twin forces which strive to maintain the status quo of the institution when it is menaced by an entrepreneur's innovation. The long-range needs of the corporation are generally disregarded in the manager's preoccupation with the time span of his own career. And the line of succession in management is geared to protect the momentum and the style of the ongoing establishment. However, as Aristotle observed, genius does not always breed genius and the sons of a great man may be fatuously dull. So it is often the case that despite pronounced policies supporting innovative change, the hierarchy is tightly closed against change, and the would-be entrepreneur is frustrated.

In another sense, it seems that an organization's innovative fertility is often in inverse ratio to its standard of living and economic development. The more successful it becomes in the economic sphere, the more precious become its momentum and its dedication to the production ethic, which has been the cause of its success. To change styles and introduce the significant changes in conduct and criteria of performance which are necessary to shift to an innovation ethic is often too much to ask or expect. As Theodore Levitt has said:

There comes a time in the life of every company when it must abandon principle and do what is right.

It has to face facts and do what vigorous survival requires, even if this involves a drastic departure from its historic product line, its historic research and development orientation, its historic manufacturing processes, its historic way of doing business.

Unless it does this it may end up keeping its pride but losing its shirt.

There are three pre-eminent facts of modern economic life that every company has to know and reckon with:
- The acceleration of change
- The growing similarity of competing products
- The growing sophistication of the customer.[1]

Innovation: Uniquely Human

It was pointed out in Chapter 1 that innovation is a uniquely human quality, for it is precisely man's ability to change that distinguishes human nature from all other aspects of nature. Other members of the animal world go through changes, but it is usually adaptive change. Nature's cycles repeat themselves endlessly, and animals change only when their instinctual life style does not suffice to guide, protect, and continue the species. Humans cause change, and as Samuel Johnson once said, "Few moments are more pleasing than those in which the mind is concerting measures for a new undertaking."

The term *to innovate* was still on probation in 1947, according to H. L. Mencken's *The American Language*, whereas in the 1963 edition of the same book it was considered to be an acceptable back-formation word. Thus it has taken almost a quarter century just to attach a word handle to this concept and put it into active use in the manager's vocabulary. Now if we can persuade managers not only to use the word but also to invoke the concept in their management philosophy, some progress can be made.

Of course, an essential part of the innovative process is the benign opposition to a proposed change which results from the desire of individuals, institutions, industries, or social systems to maintain the momentum and stability earned at a high price in energy and effort. Great productivity is accomplished with standardized designs and established methods; the point is simply that these should not be over-protected by bureaucracy. As the historians Will and Ariel Durant phrase it, "The conservative who resists change is as valuable as the radical who proposes it—perhaps as much more valuable as roots are more vital than grafts. . . . It is also good that new ideas should be compelled to go through the mill of objection, opposition, and contumely; this is the trial heat which innovations must survive before being allowed to enter the human race." [2]

Resistance to change is not confined to the establishment or to management; the innovator himself may battle ideas that are not his own. In 1899 Thomas Edison had this to say against placing electrical wires underground and using alternating current:

> There is no plea that would justify the use of high tension and alternating currents, either in a scientific or commercial sense. They are employed solely to reduce investment in copper wire and real estate.

> My personal desire would be to prohibit entirely the use of alternating currents. They are as unnecessary as they are dangerous. I can, therefore, see no justification for the introduction of a system which has no element of permanency and every element of danger to life and property.[3]

Innovation and the dark view often taken of it are by no means strictly twentieth-century phenomena. Reaching back into the murkier side of human consciousness explored in medieval witchcraft, we find Belphegor, the archdemon of ingenious discoveries and inventive and innovative acts—many of which were considered shameful. He was alleged to have cannibalistic tendencies and frequently assumed the shape of a young woman who distributed wealth. Belphegor was used by the early management savant, Machiavelli, in his writings. Perhaps we need some twentieth-century exorcisms to dispel resistance to innovation and free the modern entrepreneur. It is hoped that the modern versions of innovation have not inherited Belphegor's repulsive characteristics, although certainly some of his "feminine" mystique persists, for the concept of innovation is at once intriguing, unpredictable, potentially expensive, and possessed of intuitive attributes. There is no doubt that technological innovation has helped distribute the economic and social wealth of mankind.

Innovation: Essential for Survival

Much has been written about the manager's future tasks and his coming preoccupation with new concepts and ideas which reach far beyond the traditional focus on materials, people, finance, and technology. With the rapidly changing environment, the habitual hierarchical inertia of most organizations can no longer be viewed as a stabilizing force which avoids disruptive change. Not only does the inertia become an impediment to innovative growth when competition and the environment change quickly, but it becomes a source of formidable discouragement to the entrepreneur. Thus management style and concepts need to evolve away from rigidity toward flexibility and openness in order to adopt changes while facing new situations. Such an adjustment in attitude and philosophy will come about when the anatomy of the innovation process is better understood by the manager. This is not an easy task, for the innovation ethic is an abstract concept being applied to what is usually a pragmatic management situation.

The management challenge is to deliberately innovate from within by circumventing or containing the forces which preserve the existing corporate culture, and by ultimately transforming the culture itself. The problem is to do it with-

out completely shattering the hierarchical setup and its tremendous contribution in the form of stability and productiv-
ity-cost benefits. "To innovate is not to reform," observed
Edmund Burke in 1796. Rather, it should be to introduce useful changes adiabatically into the system, that is, without
gain or loss of the heat of the entrepreneur's fervor. Prefer-
ably it should be done without an organizational crisis, although this is often the only way to transform habitual inertia
into a readiness to try new ways and concepts.

Entreprenertia, of course, exists, down the line in the corporate hierarchy to the point of organized labor's negotiated
contractual state of stability. Rigidity, congruity, and stability
at the bottom of the corporate pyramid are understandable,
and a flexible, innovation-tolerant apex seems antithetical to
most styles of organization. But the limit to innovation, and
the real hazard to the status quo, comes only at the point
where the identity of the establishment itself is threatened
by the institutionalizing of the innovation. It is a sophisticated
and secure management that can consciously and effectively
open up its system to let innovative changes in, and then can
nurture them through their birth pains and adolescence as
the uncertainties are gradually reduced to risk situations with
which the establishment can be more comfortable.

Organizing for the Entrepreneur

The classical approach to dealing with internal inertia
which gets in the way of innovative efforts by management is
to circumvent the problem rather than solve it. In order to
spare itself the task of making the existing organization more
nearly capable of accepting an innovation, management often
examines the business sector that it is in and allows itself to
form entirely new business entities. This, however, does not
solve the basic problem for the parent entity, and it may lead
to other troubles. Where there is serious competition which
threatens the old business, management may need to change
within itself rather than to deal with change by creating sep-

arate new businesses. Conventional wisdom provides at least two answers to this internal dilemma. The first is that people cannot change and that the organization therefore cannot change unless agents of change are brought in from the outside (in the form of a new president, say, or perhaps a new research director). The second is that entreprenertia can be overcome by the use of force, which will somehow improve the internal workings of the company and lead the organization to accommodate itself to change.

However, if the problem is considered from the standpoint of encouraging the positive acceptance of change rather than eliminating resistance to it, there are a number of innovations in corporate structure that are designed to accomplish this purpose. They range from passive adaptations or mutations of the organization to its complete transformation. Acceptance of new ideas by the people in the institution is the key, of course—as it has been since the experiment in the Garden of Eden. What is new is the complexity and the rapidly changing environment that management deals with today. The major task of administrators of modern institutions is to design and operate what some scholars call *linkage mechanisms*. These mechanisms are organizational structures in nature and are integrated into the institutional system's organizational scheme, thereby modifying it. They then further integrate the modified institution with specific publics (investors, consumers, suppliers, and so on) and also obtain political and social sanction. In the end, this sanction determines the relevance of the institution, for if it does not serve society properly, it will not survive.

If the need for change originates from within, say through product or service innovations proposed by a creative researcher, or perhaps through an innovative practice of management, the usual organizational approach is to create what may be called a subsystem of the parent system to handle the change. This is because the innovation often tends by nature to conflict with and be irrelevant to the primary activities of the system.

On the other side of the change process, the traditional

course of institutional growth is through the relentless momentum of relevant "improvements" in products and services provided by the basic system. These can become irrelevant to the environment, present or future, faced by the institution, as when the system churns out a product or service the customers don't want, can't use, or reject in favor of a competitor's innovation. The manager's challenge is to steer a course between creative activities which are irrelevant on the one hand, and relevant activities which are uncreative on the other.

At the University of North Carolina, Professor Rolf P. Lynton made a survey of recent research and hypotheses concerning various integrative devices for handling change.[4] He concluded that both large and small institutions face circumstances in which they can no longer deal with change by intuitive, interpersonal, spontaneous, face-to-face contact. Attitudes are not flexible enough, and freedom within the organizational structure is insufficient to allow evolutionary change. Modern institutions must resort to some formal approach in order to cope with change.

In the past, an organization's size was considered the primary factor in deciding whether to formally reorganize for greater flexibility. An implicit assumption was that, as a system enlarges, it has more difficulty in handling change. Smaller organizations are noted for their ability to deal informally with innovations, and business history is studded with major innovative breakthroughs achieved by smaller companies. The recent academic findings described by Professor Lynton, however, indicate that the decision to formally cope with changes is not so much related to organizational size as it is to the *degrees of uncertainty* in technology, markets, and the future environment which the institution faces. The manager must redesign and transmute his organization into one which can accommodate higher degrees of uncertainty, and which can also deal with any subsystems created for the purpose of developing innovations. Examples of innovative subsystems are readily found in the chemical and electronics industries. The government's creation of

Comsat, the Peace Corps, and Medicare, to name a few, also exemplify a system's attempt to find innovative solutions to problems through specially designed subsystems.

In dealing with uncertainty, we are dealing with more data than can be handled, or with conflicting data, or in some cases with a significant lack of data. The available information is also usually located at a lower level of the organization. There it remains until the bureaucratic organization develops alternates that appear to be options, which as a rule must be made without adequate data and presented to top management for consideration. Developing the alternates requires what is called "leaps of decision" rather than the logical step-by-step program of problem solving. Under the circumstances, it is the person down the line rather than the chief executive officer who must make these leaps of decision, although in many instances he does not have the experience, maturity, or insight of a seasoned leader. Thus the organization is usually attacking the problem of uncertainty at a level that is not optimum for effectiveness.

Dominating Issues

The problem of innovation in uncertain areas cannot be considered in a vacuum; the interrelationships among the task, the environment, the resources (including technology), the people, and the structure of the organization are important.[5]

With respect to the *environment*, this discussion assumes that there is a state of uncertainty which is treacherous and unknown, and that the uncertainty must be reduced to risk before real business progress can be made. Although most good managements are skilled in problems of risk taking and risk avoidance, very few are expert at making leaps of decision in an uncertain situation.

With respect to *resources*, including technology and time, management's past emphasis has been on productivity and cost as distinct from innovation and creativity. Management has moved from the early days of concern with materials, through a period of people management, to a preoccupation with financial management. The challenge of managers in the future will be to deal with the extensive changes in technology and the vast amount of information which will become available. The management of technology, the management of time, and the management of ideas all require a different management style from the styles of the past.

In the case of *people*, usually the managers who are enthroned in an existing organization got where they are by successful performance in managing materials, people, and money. Thus their experience relates to yesterday's problems and to yesterday's environment. Their style of management is often that of risk avoidance, with the dead hand of corporate control steering the institution steadily in the same direction as in the past. They frequently have a fervent commitment to a rational setup, to traditional activities, and to the status quo, particularly as regards organizational design.

The entrepreneur and the innovator, not surprisingly, are usually unhappy with the constraints imposed by these managers. In an M.I.T. study of 44 key people who left one electronics systems company in the Boston area, it was found that they had formed 39 separate small companies, 32 of which had survived after five years. These 32 had total sales of $72 million, compared with the $30 million of the parent company from which they had fled.[6]

Structure and its interrelationship with the innovation task is affected by the fact that a company is usually organized to handle risk, not uncertainty, and is structured to handle yesterday's problems and ideas. As a relatively closed system, it resists change and maintains its own momentum; it does not readily open up and let change in.

Organizational Climate for Innovators

A study of the organizational atmosphere that encourages innovation was undertaken by Victor A. Thompson, chairman of the political science department at the University of Illinois, in 1968 when the Great Society programs were bogging down in many big federal agencies. When the investigators actually examined firms that have successful records of innovation

> They found a situation very different from the traditional model of the taut ship. They found what, in an innovative organization, is called "slack" — uncommitted resources of personnel, finance, material and motivation. Although we don't understand exactly how slack increases innovation, the fact that it does is well established. Perhaps the embarrassment of unused resources leads to the search for something worthwhile to do with them, or perhaps a use is found for them simply because they are there.
>
> When there is slack the psychological risk of new ventures is reduced. The possible loss of uncommitted resources is less painful than the loss of resources that are already earmarked for specific use. Like the rich gambler who can afford to lose, the executive who has many uncommitted resources at his disposal will play a riskier game for higher stakes than one whose resources are so limited he must play to minimize loss.[7]

Thompson points out that rich farmers are the ones who make agricultural innovations; farmers who just scrape by stick to tried methods. Progressive school systems are found mostly in wealthy suburbs, not in a city's working-class or poor neighborhoods. He notes that diverse approaches combined with a fairly loose organizational structure have also produced a high innovative payoff in industry. The firm that has made the most successful innovations in the manufacture of jet engines, Pratt & Whitney, is a pragmatic, trial-and-error decision maker. This company avoids a rigid top-down, comprehensive problem-solving style. Innovation in the aluminum industry was rare when the giant Alcoa Corporation dominated the field, but increased after 1945 when other companies broke Alcoa's monopoly.

Another study, made by William W. McKelvey, which centered on 121 employees from two divisions of a research organization, explored what happens when professionals find that the organization is not fulfilling their research and career expectations. Results showed, first, that the perception of expectational unfulfillment is highly correlated with cynicism and active expressions of disapproval; second, that cynical, actively hostile professionals received the lowest promotion eligibility rankings from their supervisors. In contrast, idealistic, passive professionals tended to receive the highest eligibility rankings.[8] The manager concerned with entrepreneurs and innovators will be faced with these problems. He can view his people as characterized in one of four ways, according to the study.[9]

The crusader, or active idealist, expresses positive sentiments and confidence that the organization is an ideal structure in which he can obtain his professional expectations. He also believes that it has some flaws in conflicting expectations, showing up in its actual behavior, that do not complement his own expectations. He responds by trying to correct the flaws: he emphasizes his expectations and continually presses to have his ideas for improvement adopted.

The insurgent, or active cynic, expresses negative, hostile sentiments and loss of confidence in the organization. He perceives it to be a structure in which he definitely cannot achieve his professional expectations, and thinks of it as having few redeeming virtues. He responds by deciding that the creation of organizational expectations which complement his own expectations cannot be achieved without destroying the present structure and building a new one.

The ritualist, or passive idealist, expresses positive sentiments and confidence in the organization as the ideal structure in which he can attain his professional expectations, but he is very much aware of some flaws and conflicting expectations appearing in the organization's normal behavior that do not complement his own expectations. In contrast to the active idealist, the passive idealist responds by conforming to the noncomplementary expectations and de-emphasizing the expectations he desires, rather than risk his position by trying to change the organization's behavior toward that of the ideal structure.

The retreatist, or passive cynic, expresses negative, hostile sentiments and loss of confidence in the organization. He perceives it to be a structure in which he definitely cannot achieve his professional expectations, and thinks of it as having few redeeming features. Like the active cynic, he responds by deciding that the creation of organizational expectations which complement his own expectations cannot be achieved within the present structure. The passive cynic, however, withdraws rather than advocating the organization's destruction.

McKelvey's findings can help managers recognize that the professional's value system is distinct from the set organizational posture assumed by conventional institutions. The entrepreneur is a professional of sorts, certainly in the sense that he has his own value system, career objectives, intellectual code (implicit or explicit), achievement motives, and power drives, which often are antithetical to the organization in which he works or which is supporting him.

Organizational Models: Classical Patterns

There are at least five basic patterns that must be dealt with in existing organizations when management considers means of stirring the innovative spirits within them. The classical organizational designs below the CEO (Chief Executive Officer) are as follows.

1. *The traditional functional organization* with the standard divisions of manufacturing, marketing, finance, R&D, personnel, and so forth. (Exhibit 1.)

2. *The product type of organization,* where the company groups its product divisions or departments into separate organizational profit centers. (Exhibit 2.)

3. *The business grouping,* in which the company has grown in size to the point where it organizes by division,

subsidiaries, or departments that operate separate businesses. (Exhibit 3.)

4. *The geographical grouping,* where the company, which is usually in a single type of business, is organized by geographic region, with operational and profit-and-loss centers for different locations around the world. (Exhibit 4.)

5. *The individual-leader organization,* where an outstanding individual sets the style and the manner of operation.

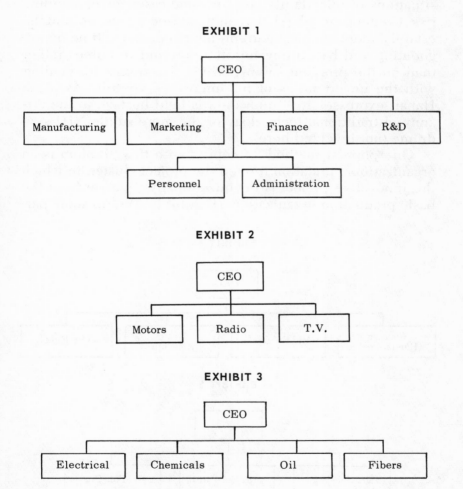

EXHIBIT 1

CEO

Manufacturing | Marketing | Finance | R&D

Personnel | Administration

EXHIBIT 2

CEO

Motors | Radio | T.V.

EXHIBIT 3

CEO

Electrical | Chemicals | Oil | Fibers

This may be a mixture of other types of organizational design, implicit or explicit. (Exhibit 5.)

In each of the five classical forms of organization, the employees inevitably develop an emotional and intellectual investment in maintaining the established design. This investment is a built-in resource which is advantageous to the existing business and its style. In time there accumulates a managerial bias toward maintaining the status quo, avoiding situations of uncertainty, and in some cases, even avoiding risk. Conflicts develop between would-be innovators and the establishment, and the entrepreneur spends his time on negotiating and bargaining for attention and resources rather than on the problem solving which is essential for dealing with the uncertainties of his innovative venture. As Mack Hanan expresses it: "Innovation is held hostage to the tyranny of traditional technology or the more subtle NIH syndrome (not invented here)." [10]

One popular method for coping with this problem is an organizational mutation termed the project cluster, in which the innovative project is organizationally separate from the basic production organization. (Exhibit 6.) Within some par-

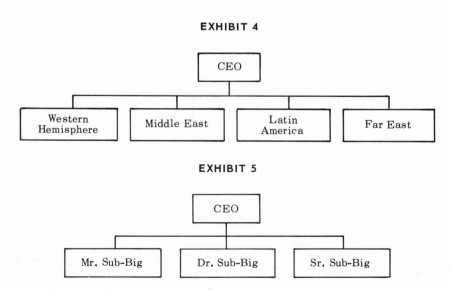

EXHIBIT 4

CEO

| Western Hemisphere | Middle East | Latin America | Far East |

EXHIBIT 5

CEO

| Mr. Sub-Big | Dr. Sub-Big | Sr. Sub-Big |

EXHIBIT 6

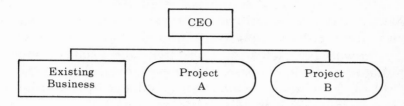

ent organizations it becomes a subsystem in the form of a special task force; a special assignment of a product champion, a product manager, or a product administrator; or even a special department or division set up (often on a temporary basis) with a given time frame for accomplishing the innovative development.

In its more recent form, the project cluster is set up outside the parent organization as a permanent subsidiary, sometimes 100 percent owned by the parent, sometimes under public ownership, sometimes with equity participation by key personnel. The primary characteristic of this external innovative system is that it is designed to encourage creativity and innovation without coming under the influence of the productivity theme of the parent. In its initial stages, it has the advantages of rapid communications and a circular (committee) organization rather than the usual star-form, hierarchical one. Its structure is fluid, job definitions are flexible, it encourages vertical and horizontal teamwork, and usually it is organized with multidiscipline people.

Some form of link must be developed between the parent organization and the subsidiary which will allow the innovative system freedom, yet provide it with ready resources, some guidance, and some intelligent integration with the overall institutional scheme. This interface between the parent institution and the innovative subsidiary is critical. Ultimately the innovative subsystem must be sanctioned not only by the parent-owner, but by the customers that it serves and by its outside owners, if shares are held outside or by the participants. If this sanction is not wholehearted, the innovative subsystem cannot survive despite its organizational design.

The innovative subsystem is designed to provide enough leeway so that the intuitive "leaps of decision" can be made without the arduous time-consuming drill of getting a series of approvals for each use of resources and each movement into a new or uncertain area. However, even the innovative child of the parent organization is subject to the propose-dispose dynamism: it must have its propositions approved by the hierarchy of the parent system, unless arrangements are made to put such decisions into an entirely different frame of reference and control. Unless this problem is recognized and solved, innovative subsystems are peopled by what David Gleicher of ADL calls "entrepreneurs without authority."

In considering structures for encouraging innovation, given an existing establishment, the cost and the benefits of each structural setup must be considered. In the five traditional patterns described earlier, the existing organization has been expected to develop innovations, and in many cases it has. However, a major change in attitudes toward innovation has not often been accomplished within an existing organization. In the five classical models, innovations are expected to flower and mature under an operating business which is set up to improve productivity. It is a rare operating manager who can come through with a major innovation in his system; entrepreneurs are usually missing. Such a structure can be characterized as a closed system dedicated to productivity, its concern being with minimizing risks rather than dealing with uncertainty. Major innovative efforts are considered a diversion. The greatest benefits are derived from maintaining the status quo, and innovative thrusts are considered disruptive.

Many companies have decided that the research and development arm of the company can provide a better climate for nurturing an innovative effort than the operating departments. When innovations are focused in the R&D area, however, several problems arise. For one, the NIH (not invented here) factor flourishes in the R&D setup. For another, innovations that have been assigned to an R&D organization have been delegated down the line where less business experience is available. Further, the operating departments may consider the innovative efforts a threat to their own positions. Perhaps most important, there is no profit-center atmosphere for the innovation when it is lodged in the usual R&D function.

As already mentioned, the method of dealing with innovation via the project cluster presents some problems too. When the parent system creates a separate department for all new enterprises, or a new-ventures division as a major new profit center, the enterprise group can become entrepreneurs without authority, forced to bargain for resources and compete for status. Although this situation can be avoided by making the innovative system an entirely separate organization, with a different decision maker who reports only to the parent's board of directors, problems of corporate identity may arise between the subsidiary and the parent. If the board does its job, the innovative system's frame of reference, control, and planning are entirely unrelated to the system of the parent. Motivations are different, for this is a separate profit center in a separate business. In the case of public ownership, the responsibility of the trustees to their public shareowners causes decisions to be made in the best interest of the innovating organization's objectives, which may be unrelated to the parent's objectives.

Organizational Models: Innovative Evolution

The following schematic diagrams trace the evolution of efforts by ongoing institutions to create innovative organizational entities within and without their basic system. These patterns are derived in part from various academic concepts on linkage in innovative subsystems.[11] The code used in the drawings is —

D = decision maker
O = operator of existing activity
I = innovator linkage
E = entrepreneur

The five traditional organizational models described previously — the functional, the product, the business grouping, the geographical grouping, and the individual-leader types of organization — all tend to resist innovation by using Pattern 1 (Exhibit 7) or Pattern 2 (Exhibit 8). As already discussed in this chapter, Pattern 1 has many disadvantages. The system tends to be closed and stabilized within an established perimeter. Entrepreneurial activity is an interruption to operations if the entrepreneur reports to the operating head, or to the decision maker if he reports directly. The system is set up to deal with certainty, not uncertainty, and to avoid risk. Moreover, the real information and entrepreneurial know-how is focused at a low organizational level rather than topside. Since the system is dedicated to productivity instead of to innovation, the decision maker has no transcending commitment to innovation, and the entrepreneur tends to disrupt the status quo.

Pattern 2 (Exhibit 8), a mutation of Pattern 1, is sometimes set up on the assumption that the research and development function is more sympathetic to entrepreneurial effort than

EXHIBIT 7

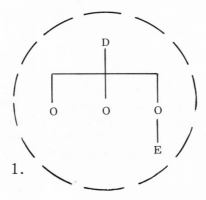

EXHIBIT 8

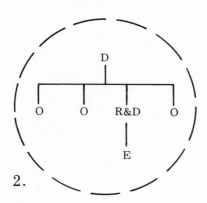

2.

the operating departments. Drawbacks of this pattern include the NIH factor, which rises in the operating sectors of the organization, and the delegation of the field of uncertainty in which the entrepreneur exists to too low a level for dealing with it effectively. The research and development function is often the least experienced in the business management which is required to make a successful innovation. The entrepreneur presents a political threat to the other sectors of the organization in terms of allocation of resources, advancement, status, and the like. Also, the research and development sector of the institution is usually the least profit-conscious, because R&D by its very nature is not normally a profit center in conventional accounting terms.

Pattern 3 (Exhibit 9), the project cluster, aims at a better balance of the cost-benefit equation by providing a linkage mechanism. This is in the form of an innovator responsible for moving the entrepreneurial activity to a commercial or useful stage from its creative beginning in an area of uncertainty. Although still a closed system, the model permits a champion to be identified in an explicit organizational sense. It gives him a flexible communicating setup direct with individual entrepreneurial foci and relieves him of some hierarchical constraints, for his unit becomes a profit center equal in rank to the operating profit entities. However, he is

EXHIBIT 9

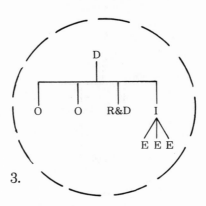

3.

handicapped by the need to bargain competitively against operating entities for the allocation of resources, so that he cannot devote his major effort to problem solving. Thus a conflict structure is established and encouraged under the decision maker. The situation creates an entrepreneur-innovator without authority, unless there is an unusual delegation of power and responsibility, together with sufficient resources to accomplish the venture. Of course, the accountability for the innovative activity, as with any accountability, cannot be delegated.

Pattern 4 (Exhibit 10), a modification of the project cluster, still uses a closed system but features a task-force subsystem headed by the innovator. As advisors and participants in decision making, he has the system decision maker, one or more operating heads, and one or more entrepreneurs, who form a committee type of organization. Advantages include the creation of a separate profit center and the involvement of the decision maker, the operators, and the entrepreneurs. The information exchange and organizational flexibility which were the strong points of Pattern 3 are retained. But so are the disadvantages inherent in a single system: there is no independent association as long as the decision maker is a part of the task force, and there is hierarchical restraint because of the dual roles of the advisors. However, pattern

4 is an improvement because the base system remains essentially a spectator, and control and planning are mixed.

In a more advanced mutation, Pattern 5 (Exhibit 11) makes the project cluster into a separate institution. This involves a change in stewardship to a director-trustee relationship, as well as an entirely different frame of reference for objectives, planning, organizing, and controlling. As discussed earlier, different motivations are involved in these different profit- or service-oriented institutions. Public ownership, the ultimate step, introduces questions of discipline and other considerations. The institutional identity

EXHIBIT 10

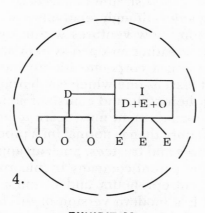

4.

EXHIBIT 11

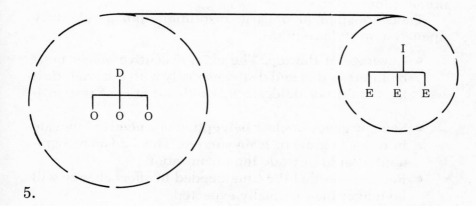

5.

of the new enterprise may even be entirely separate, if this is appropriate. Such a spin-out from an established enterprise releases the new venture from vested interests and historical expectations, but it presents a problem when the trustees and shareowners of the originating institution see the venture, with its growth and profit potential, moving away from them. This potential loss can be reduced by an arrangement that allows the parent institution to remain an investor in the new enterprise, although it is hoped that the established firm will keep its management fingers out of the venture. One successful venture capitalist's advice to executives of going concerns is to "invest in but do not manage" the entrepreneur-innovator if you wish him to succeed.

Certain companies, though relatively few, have been successful in spawning new ventures within their own institutional system by creating an open-system atmosphere. This involves developing a corporate identity as an innovative institution, or at least as one which can harbor entrepreneurs without stifling their style and course of action. The institution and its top management must learn to be more comfortable with uncertainty than is normal for a production-minded group of executives and trustees. Such an approach recreates the genesis of the parent company by incorporating zealous groups around entrepreneurs and common objectives. In some senses, it is a modern version of the family business, with the head of the family's position occupied by the entrepreneur-innovator.

Transformation of a basic institution into a relatively open-system style requires:

- Leverage at the top. The chief executive officer must be interested in and deal personally with uncertainties; he should not delegate this difficult area of responsibility.
- Some urgency or clear perception of a need to innovate in order to grow or even survive. This "crisis recognition" should pervade the organization.
- Recognition that the time needed to effect change will be longer than normally expected.

- Charismatic leadership with some readily apparent goal or vision to serve as a model for the direction of change.

The institutionalizing of innovative ventures, thereby separating them from the parent, may have the ultimate result of converting the parent into an innovative institution. 3M is a good example of an organization which started out as an innovative company and has maintained this reputation. In the case of an existing company which is known for its productivity, the gradual transformation of the basic institution into one known for innovation can perhaps be accomplished over a long period, depending on the determination and capability of the management.

The Venture Organizational Concept

Broadly speaking, the venture concept is an attempt to make innovation per se more predictable and less random. Venture funding represents a relatively new corporate approach to managing business development as a continuing commitment rather than a series of sporadic crash programs. Properly managed, a new venture can achieve significant innovative breakthroughs into new business areas, rather than simply making the incremental renovations that are typically associated with new product species in the same business. The venture concept and its capital requirements, therefore, should not compete with corporate R&D or product development. Neither should it interfere with the opportunistic, spontaneous generation of products or processes.

American Telephone & Telegraph, Dow, Westinghouse, Monsanto, Union Carbide, Du Pont, E.G.&G., and Bolt, Beranek and Newman, in addition to the cases discussed in Chapter 3, are experimenting organizationally with new venture internal and external organizations. Many other financial institutions and conglomerates are following the same route. But in spite of the impressive list of companies involved, venture management is still a virginal art and science.

Mack Hanan cites six advantages which are inherent in venturing in a semiautonomous context.

1. *A venture is unidirectional.* It is chartered for a single purpose; it always knows what business it's in.
2. *A venture is multi-disciplinary.* It contains representative skills throughout the organization.
3. *A venture is eclectic* (selective). It enjoys relative freedom in probing market needs that offer business opportunities within a broadly interpreted definition of harmony with corporate objectives. Its tendency to innovation is unimpeded by being held hostage to what is often the tyranny of traditional technologies or the more subtle inhibitions of the NIH (not invented here) syndrome, both of which regularly afflict product development departments and market research functions.
4. *A venture is entrepreneurial.* Venture teams are business-managed according to a charter and a written plan.
5. *A venture is judicious.* The venture team tries to be influenced only by facts, rejects too early emotional commitment ideas just as it resists embracing assumptions about their marketing inevitability.
6. *A venture is kinetic.* It is dedicated to change which becomes expected of it, standing ready to fill new needs. Its justification lies in doing something innovative.[12]

In Hanan's analysis, the company gets going on a venture by a four-phase path of decision making: *first,* venture selection, which is largely assumptive; *second,* information gathering, which is entirely factual; *third,* evaluation, which includes generation, formulation, and testing of new product concepts; *fourth,* recommendation for venture commercialization (that is, innovation) based on a market entry plan.

Composed of a small number of representatives from the disciplines of marketing, science, technology, and finance, a venture team seeks out products and services to meet the next-generation marketing needs of the corporate profit cycle. Venture teams can operate within their parent companies as departments (though this is likely to be difficult), as joint enterprises with other venture teams, or as independent companies themselves, if their parent corporation has adopted an innovation-spawning philosophy. So far, the effect on current corporate organization has been mainly evolutionary, with liberalizing innovation efforts under entrepreneurial business-management cells.

Among management philosophers the belief is growing that venturing's most significant value today is its influence on the classical corporate organization format and the allied concept of the corporate manager. The educational influence of venture teaming is beginning to make itself felt within the corporate structure as a method for managing existing business. Trouble lies ahead, however, unless some philosophies and styles of management are changed. Small, market-oriented teams are proliferating, each one led by a "general manager" who in theory is not just profit-accountable but profit-responsible, and who accordingly has free call on the universe of corporate and noncorporate technical, manufacturing, and marketing services. Where this approach is made to work, it is a giant leap forward; but many efforts are running into entreprenertia.

The generic concept of the venture team is being brought home to more and more top managements, not only as venture management which innovates its parent company's new-business development, but (and perhaps more important) as the frame for the question, "What will our corporate organization of the next generation be like?" The venture concept will increasingly affect the restructuring of corporate life styles.

7

Chief Executive's Role in an Innovative Enterprise

There are few situations in life which cannot honestly be settled, and without loss of time, by suicide, by a bag of gold, or by thrusting a despised antagonist over a precipice on a dark night. *Chinese proverb*

THESE three options for dealing with life's problems no longer hold even in their more civilized versions. Not everyone in management, however, recognizes that power alone is no longer a universal panacea, and certainly not when the problems concern innovation. The fact is that the chief executive's powerful directive role is fast changing. His primary task used to be to command better performance through increased productivity in the face of risks; now his role also demands systems uniqueness and a coalition with the entrepreneurial innovators who thrive on change and uncertainty.

The chief executive officer of a small venture has less trouble perceiving his contemporary role as a quick-change artist than the CEO of a large company trying to launch a new venture, whether inside or outside the going business. The entrepreneur-innovator standing alone has no heritage of the production ethic to contend with. But in an established producing entity, there is a tendency for the innovator to be swept into the basic institution's bureaucratic setup. The simplistic power-driven bureaucratic style must give way to an entrepreneurial teleological style if an enterprise is to be innovative. The bureaucratic style, based on the rule of law, is primarily focused on internal affairs, productivity, and efficiency. The teleological style, based on rule by man, involves the release of human energy toward a given end. It focuses more on external effects, and has an economic dimension of investment rather than cost.

The challenge for the chief executive, whether he heads a newborn venture or an established company seeking an innovative dimension, is to develop a personal style which takes the best from the bureaucratic and teleological worlds. In neither case can accountability for innovation be delegated by the CEO, any more than the accountability for production can. Authority for production, however, can be delegated, provided that it is properly coordinated, that organizational assignments are clearly set forth, and that a management information system is established to keep the CEO informed. Authority for innovation, in contrast, cannot be delegated with such confidènce, because innovation is intrinsically fluid, is concerned with the untried, and is involved in areas where new policies must be formulated and where old policies often inhibit entrepreneurial innovation.

CEO's Look in the Mirror

Innovation is our motto; it's just not being practiced! *Anonymous*

Paraphrasing Mark Twain, to be innovative is good; to teach others to be innovative is even better and much easier.

Perhaps executives can learn something about the innovative approach from a study of the CEO role in which a group of international executives viewed their current situations. While the study was designed to explore all types of enterprises, it did yield some results that are germane to the CEO seeking a more innovative role.

Conducted in 1969 by the Conference Board, this study described the job of the chief executive as unique in the sense that he has ultimate responsibility.[1] It suggested that in no other position does the character of the individual so completely mold the nature of the job. According to interviews with more than 300 chief executives from all continents, the meaning of the term *ultimate responsibility* rests on three points:

1. Responsibility for growth, as evidenced primarily by financial performance.
2. Responsibility for the company's character, as evidenced by the reputation it has earned.
3. Responsibility for perpetuation, as evidenced by the company's capacity for self-renewal and sustained momentum.

It is in the third point of perpetuation that the increasingly recognized innovation mandate emerges. It is related to the CEO's responsibility for the character of the company in that a reputation for being an innovative organization is vital in attracting certain types of employees and investors. And a record of innovation results, it is hoped, in an outstanding growth record, so that the responsibility for innovativeness is woven equally through the three points of ultimate responsibility of the chief executive officer. Thus innovation and the management of change are a major and pervasive responsibility of the chief executive: this is the innovation ethic.

As for the CEO's decision on what to delegate, the Conference Board study showed that four factors are involved:

1. Confidence in subordinates.
2. Time available.

3. Impact of decision on the company's future.
4. Novelty of the situation.

It is the fourth point which indicates the chief executive's wariness about delegating in areas where the company has had insufficient experience to form guidelines—that is, in areas of uncertainty. This is usually where innovation takes place. Perhaps a fifth factor should be added to the list: the CEO's confidence in himself, particularly in his ability to innovate.

An executive understandably is more hesitant to delegate in a situation that is too fluid or uncertain to lend itself to guidelines, principles, or parameters drawn from his or the company's experience. Just what constitutes an uncharted area is not clear, since it varies widely by company and CEO. It usually includes such tasks as the selection of senior key executives and negotiations regarding mergers and acquisitions. But as one executive explained, even this kind of responsibility gets delegated if he believes anyone else can handle it.

According to the Conference Board study, the CEO perceives a distinction between the role of manager and the role of leader. The respondents were asked whether they considered themselves primarily owner-managers, professional managers, or entrepreneurs, although these terms were not specifically defined. About 10 percent saw themselves in two or three roles, and saw these roles as shifting with the varying demands of the job.

The head of a small company in Latin America said: "I was an owner-manager, but as the business grew I had to move to become more of a professional manager." A CEO in the United States who had been hired to head a long-established company saw himself primarily as an entrepreneur, with innovation as his job. Well over half of both U.S. and non-U.S. chief executives thought of themselves as professional managers. Even so, however, "there seems to be little justification for cheer or moan that the entrepreneur is dead, long live the professional manager." Rather, it is evident that searching out new risk opportunities, innovating, and developing new combinations of resources will continue to be a major element in the chief executive's duties, although obviously not the whole job.

From this study, two broad comparisons can be made between U.S. and non-U.S. managers as they perceive themselves with respect to the entrepreneurial-innovation function: (1) Seven out of ten U.S. chief executives consider themselves to be professional managers, whereas only six out of ten non-U.S. chief executives do. (2) Twice as many non-U.S. managers consider themselves to be entrepreneurs as U.S. managers do. While this amounts to only two out of ten managers outside the United States, only one out of ten within the United States sees himself as an entrepreneur.

Some people assume that CEO's themselves must be and in fact are creative, and that established organizations generally desire to inculcate and nurture an entrepreneurial spirit. This is not true of all top managers or of all large organizations. To those with such an objective in mind, serious impediments arise. Many large industrial organizations have reached the conclusion that a low yield of creative ideas can be expected to come from a top management saddled with the pressures of operating a business. The environmental and style changes necessary to stimulate creativity are often antithetical to conventional operating modes. Other approaches are therefore necessary.

One solution which was explored in Chapter 6 involves establishing an entrepreneurial cadre in a separate organization reporting to top management. The organization is independent conceptually and perhaps statutorily of the operating divisions. If it is to have much chance of success, it will be politically and emotionally free, with patient, quick, and ready support from a top management which is attuned to the "bank rolling" type of venture financing rather than to traditional criteria such as ROI over a long period.

Stand-alone systems may sometimes be the best way to encourage disparate creative activities. Autonomous subsidiaries can be formed with few ties to the larger corporate apron strings, perhaps utilizing holding-company corporate architecture with flexible financial control. This "greenhouse" approach of establishing a separate but somewhat controlled environment allows new creations to grow to a stage where they can be transplanted to a setting more appropriate for commercial development and eventual public ownership.

The Innovation Flywheel

When an organization is well established and successfully accomplishing its traditional production functions, it is not surprising that the chief executive finds it difficult to divert attention, resources, and support to changing the company style to an innovative one. Six times as much power is needed to start a flywheel from a standstill as to keep it going once it is in motion. The CEO must expend something like six times as much effort to get his innovation flywheel started as to keep the existing enterprise moving in its traditional directions.

Some authorities have proposed that perception of a crisis is necessary in the basic institution to make this change, that it is not sufficient for the chief executive officer to proclaim: "We now plan to be innovative." Although it is true that the organization itself must feel sufficiently interested to welcome an alteration in its course, few CEO's or institutions recognize that starting the innovation flywheel requires a quantum increase in effort.

It is questionable whether a traditionally successful company can afford to have a major revolution on behalf of innovation. It would appear preferable for the leader to have a touch of charisma and a teleological outlook—for him to offer a concept and a specific model so that the direction of change will be clear to everyone. He must be definite—"Let's go in this direction in order to accomplish this purpose"—and he must have the conviction and the leadership to accomplish his objectives. The process of setting these objectives has been described variously as "charting the corporate destiny," "gap analysis," "new-ventures analysis," "new-enterprise design," "resolving the identity crisis," and so on. Following is a commonsense approach used by many consultants and managers in determining the direction and rate of change for an existing organization.

First, determine the overall objectives of the firm. This can be the most difficult task, for many successful companies and executives abhor the explicit objective. Stated goals should include a quantitative target in various terms: earnings per share, return on investment, cash flow, sales growth, and so on, depending on the longer-term aspects of the plan and on whether acquisition, merger, or other such objectives are involved. Contributing to these targets would be earnings expected from the continued growth of the existing business, from some additional markets developed for the business, and from an innovation component over and above what will come from the business. Objectives must be attainable, and they must be related to the inner desire of the chief executive officer, who will be accountable for attaining them.

Second, realistically analyze the present state of affairs in the company and its capability. This includes a sober look at the present business, its market development possibilities, and the resources that are available. Quite often this rigorous analysis is difficult for people within the company to make, and outside assistance is warranted.

Third, calculate the differences between the firm's objectives and the analysis of its actual expectations. (Dr. Michael J. Kami suggests that, as a rule of thumb, it is wise to subtract 25 percent from actual expectations for overestimating "business as usual."[2] Competition, environmental changes, and other factors are generally not given adequate consideration.)

Fourth, set up a growth plan and a procedure for growth along the following lines.

Establish a separate organizational entity: an innovative division or a subsidiary with a special permanent assignment to plan and accomplish new-venture objectives. Dr. Kami suggests that, in considering new ventures, a 35 percent success factor might be appropriate. The present writer's experience is that this is too high an expectation in some business areas — that 15 to 20 percent may be a more useful guide.

Motivate the existing organization to attempt innovations in current product lines, processes, and distribution systems. With effort, "garden variety" innovations can be developed to renew some of the existing business. To expect that all innovations should come from a separate task force or subsidiary is to miss a major opportunity for innovative improvements.

Set up a faster decision-making process for dealing with innovative efforts in the separate units and the existing organization.

Design a faster feedback and control mechanism, with a frame of reference appropriate for innovative businesses. Such a frame is usually different from that of the existing business and must be shaped as the new venture progresses.

Explore all overtures and directions for new ventures, without making judgments before all facts are available. Mergers, acquisitions, joint ventures, new products, and new services can originate from within the organization as well as from without. A screening effort should be made by personnel who are skilled in new-development work, who are not burdened with responsibility for keeping the production flywheel going, and whose career depends on starting up the innovation flywheel.

Two Views of Management

Many executives were schooled in the early days of industrial engineering, and their careers flowered in a period of rapid technological innovation and industrial expansion. Many bureaucratically trained executives tend to have a security-minded, Horatio Alger-like background, and most executives today got where they are by climbing the traditional ladder of management.

This is a frustrating career path for the entrepreneur, who wants to drive directly toward his innovation goal. He cannot be bothered with the establishment's bureaucratic impedance and traditional ladder. Entrepreneurs often have a gambler's daring along with their creativity, and unfortunately, many possess little business experience or personal concern for security.

The gulf between these two career forces needs to be closed in the interests of the growth, indeed survival, of the many established organizations which seek (or should seek) innovative dimensions to their business. The gulf is the reason that so many entrepreneurs spin off and start their own companies. Business is becoming too complex for top management to ignore any talent, either entrepreneurial or bureaucratic, that is available. The challenge of managing change in the larger companies is complicated because of the entreprenertia discussed in Chapter 6. More chief executives of traditional firms must learn to think the way the entrepreneur thinks, and more entrepreneurs either must learn to operate as well as the traditional executives do or must relinquish the straight managing function to professional managers. Business and industry have already profited by the production ethic; now they have a mandate to cash in on the innovation ethic—the art and science of managing entrepreneurial change.

In examining the current management of innovation, we find places where managers look good and places where they need to look better. Established executives usually do well in developing physical production, giving economic support, and applying known technology, but they tend to give only lip service when it comes to allowing change into the system.

Where the CEO often has trouble is in developing clear-cut objectives for his business, with a charter, a conceptual approach to growth, and true corporate and personal commitment which is communicated adequately within the organization. This program involves formulating alternate management strategies and methods to motivate and nurture entrepreneurs. A particular need for improvement is in techniques of assessing new ventures and innovations — singling out the ones to cut off soon and the ones to push harder, in order to be more selective in the allocation of resources. An example was set by the imperious connoisseurship of Lionel Nathan de Rothschild as he directed his 200 gardeners at Pavillac in selecting and nurturing the best prospects from more than 1,200 rhododendron hybrids. Rothschild gave each plant as much care as quality required, yet stringently eliminated those seedlings that flowered but were not of the best. He was ruthless with his burnings, even after 10 years' cultivation, so that he would not have even one flower among the hundreds of thousands which was not superb. His son Edmund has carried on this purgative tradition, not only in his gardens but also in his business.

Accepting all the imperatives of the innovation ethic is a large order for the CEO. One practical way to approach it is to select some significant issues involved in the management of entrepreneurial change and work at them. Among the key problems to be dealt with in developing a coalition and synergistic effort between the more conventional and comfortable production ethic and the innovation ethic are the following.

1. Corporate and self-interest in the existing momentum of the established enterprise.
2. Organizational dynamics: home office commitments, the attitude of the board and the chief executive officer, and the political-economic environment within the company.
3. The variety of management styles inherited and challenged by the entrepreneur-managers concerned with innovation.
4. The folklore concerning entrepreneurs, inventors, and innovators.
5. The demand for a new philosophy supporting an open rather than a closed system. The need to learn how to allow change to come in, and how to deal with a state of constant conflict and uncertainty.

6. Potential long-term business conflicts between various innovative enterprises; judging the timing, development, and allocation of resources to each effort.
7. International competition and the individual styles involved.
8. External forces, such as government influence, public relations, taxes, and ecology.

Managements are improving in social and cultural awareness, and it is suggested that they also need to become more aware of the teleological approach to the management of change. The process of innovation should be recognized as a part of the overall management function, which requires the release of human energy toward an innovative goal — not simply the containment of energy toward a productivity goal. The job enrichment movement espoused by Professor Frederick Herzberg, American behavioral psychologist, dismisses the industrial "mass bribery" approach of greater income and benefits as motivating carrots concerned with more "job hygiene." He maintains that this approach overlooks man's other needs to achieve, and through achievement, to experience psychological growth. While Professor Herzberg suggests that we let Mickey Mouse people who do not want job enrichment do the Mickey Mouse jobs, the task of motivating people who do want to grow or innovate is dependent upon making the job worthwhile in itself. Such an undertaking involves job content and job environment — which are in large measure under the influence of a chief executive officer seeking an innovative dimension to his business.

PART TWO

International Development and the Venture Capital Movement

8

Innovative Start-ups

In conversation, listen to what is being said;
in starting an activity, watch what is being
done. *Marcus Aurelius*

An innovative start-up, including the initial formation of a
new venture or a change in the existing enterprise, is dependent upon the environment and the setting in which it is
attempted. The impact of technological and social innovations on the international scene today is resounding, and one
measure in international development is the index of successful changes made by the public and private sectors to improve social and economic welfare.

Part One of this book described the many difficulties involved in accomplishing successful innovation within an
established organization: a manufacturing or service enterprise, a government, a philanthropic institution, or an educational system. But this was innovation in an ordered, stabilized society; innovation in a developing, relatively unstabilized society has its own set of priorities and problems.
In both situations, the difficulty is not so much to initiate the
change as it is to build and support the change long enough
for it to become accepted and institutionalized within (or
without) the system.

The process of innovating in well-established societies, and the great progress made there in introducing useful changes, products, and services, has formed a model for the developing countries. However, the forces at work are particularly complex in the developing nations, where intellectual and monetary capital resources are usually in shorter supply than natural resources.

Part Two of this book concerns innovative activity that must be started either from scratch or close to it, whether it is in a developed or developing economy. The important common denominator — venture capital — is discussed in detail in Chapter 9. This chapter will deal with the forces at work which tend either to aid innovative growth or to restrain it, and with the factors which are critical to the emergence of a successful innovation.

Forces at Work

Economic, social, political, technological, and financial forces all have an influence on new-venture formation, and an even more important effect on its subsequent development into a successful innovative enterprise or social innovation.

Economic forces at work in a newly emerging economy include the programs for developing an infrastructure of transport, communications, utilities, defense, and production which grows out of national government-financed research and development. All these public efforts create new-business opportunities for products and services. There is also an opportunity for technological transfer from government-sponsored work to large consumer and industrial markets outside the region. Other factors which affect a developing country's economic climate for innovation include the attitude of the government and of the populace toward economic growth per se, the national goals in terms of higher standards of living, and the encouragement or discouragement given foreign capital.

Social and cultural differences are often marked between innovative countries and those that are less innovative. The traditions of a society and the nature of the market provide much of the motivation for an entrepreneur-innovator to get a new venture started. Does the society value the profit motive as well as the philanthropic-service motive? Does it tend to reward or to punish innovative efforts? Hard work, long hours, and dedication to change and to new venture may—or may not—be considered a virtue, depending on the cultural milieu. Job mobility and absence of centralized government planning give degrees of freedom that are also important in certain societies.

Political forces that create an innovative climate include the presence of scientific and technological know-how, which arises principally from (1) government-funded research and development institutions, (2) private universities and research institutions, and (3) investors with technical expertise in addition to money for investment. Favorable or unfavorable tax structures and financial regulatory conventions are controlled through the political system, as is the vitally important educational system.

Technological forces vary widely in scope and sophistication throughout the world. Science and technology, however, know no boundaries; they are available everywhere and can be rapidly assimilated in the free world, providing there is the ability to receive and develop them. This, of course, depends upon a country's economic viability, the cultural attitudes of its people, and the state of its educational system. In the less developed areas, where ability to receive more sophisticated technology lags behind that of other nations, most governments have mounted a major attack on the problem through education, manpower training, and development planning.

Financial forces concern such matters as the existence of sources of capital and the attitude in the country toward capitalism. The factors involved include (1) practices of providing access to land, labor, money, and managerial talent for effecting changes and encouraging new ventures, (2) attitudes of banks, foundations, funds, and private investors, (3) existence of a public stock market, (4) financial support by the government for specific new small-business ventures, and (5) the sophistication of the financial institutions in organizing venture capital companies.

In the less developed areas, competition by large enterprises, the economy itself, and supporting and constraining factors in the environment have not yet formed the ordered pattern typical of industrially advanced countries. A major consideration of the emerging nations is their transcendent need for traditional products, goods, facilities, and services rather than for the sophisticated, high-technology market items that are the innovator's specialty. Building up a country's economic infrastructure, banking system, basic industries, and educational facilities requires development efforts, development capital, and development types of managers rather than entrepreneurs and innovators. This stage of development, in which an economy is receiving considerable government support as well as outside investment help, is in a sense a precursor of the more innovative, high-technology ventures which were the focus in Part One.

The Innovative Brew: Major Ingredients

As a Frenchman has pointed out, innovation is in some ways like love: it cannot be totally planned. Part One emphasized the fact that innovation is a uniquely human activity and described it as a process by which an invention or an idea is translated into the economy for use. In other words, while Benjamin Franklin was the discoverer of electricity, the fellow who invented the meter was the one who made all the money. In these days he is the person we would call the innovator.

The primary ingredients which coalesce or interact to form the innovative brew are first, an individual—the man; second, a concept—the idea; and third, an environment—the culture or milieu. If these ingredients of the creative process are right, they can spawn innovative activity, regardless of the area's state of economic development. However, moving a creative start to the stage where it can be classified as an entrepreneurial innovation involves the entrepreneurial gathering of necessary resources. In less developed economies the available intellectual and monetary resources are often quite limited, and the successful innovation of a product or a service useful to the society presents problems entirely different from innovative problems in a stabilized, developed environment.

The marriage of resources—intellectual, technological, financial—with a given market is essential in the process of innovation. Innovation is not an idiosyncratic accident. It comes about when through no set pattern, the major ingredients—the man, the idea, and the environment—obtain sufficient financial support, satisfy a market need, and receive public sanction.

Certain well-publicized centers of entrepreneurial innovative activities in the United States exemplify this marriage of resources. The Golden Circle—Route 128, comprising more than 700 different companies situated generally near a highway ringing the outskirts of Boston—has been chronicled for over twenty years. It was given even more publicity as a "new Appalachia" during the early 1970s, when cutbacks in defense and government expenditures dealt a severe economic blow to many of these high-technology companies. There was then a scramble to divert their assembled resources to civilian goods and services and to shunt their research and development efforts into more socially oriented areas.

Other regions in the United States—such as the Palo Alto area near the Stanford campus, the Research Triangle in North Carolina, and the area around the NASA installation near Houston—are examples of geographical concentrations of entrepreneurs, innovators, new enterprises, and new institutions. These entities, which attract venture capital and brainpower, are usually technologically based and government-related. Originally their market motivation concerned products and services stemming mainly from advances in the physical sciences and technology. More recently, however, social needs have been getting a boost from public funding of social science research and development programs in fields such as environmental management, education, and consumerism. The inputs from independent entrepreneurs, university faculty members, contract research and consulting firms, and government-sponsored research and development institutions are key ingredients of an innovative brew which has made the subject of innovation one of the most important in the world today.

Other countries are studying the possibility of replicating these innovative areas abroad. Israel, France, and the United Kingdom, for example, have had teams of government, business, and university people analyze the regional innovative phenomena in the United States. The purpose of this discussion, however, is not to detail any replicative developments but rather to highlight the differences between the innovation movement with ongoing institutions in stabilized environments, such as those discussed in Part One, and the newly created enterprises which often start from a zero base in a relatively unstabilized setting in both developed and developing countries.

The barriers to innovation are almost never technological but rather social, political, financial, economic, or a combination of these. An entrepreneurial enterprise is often the antithesis of planning, and like Topsy it just grows, if it can. The primary barriers to such innovation in any society are (1) the structure of the society itself and the state of business, political, and educational institutions, (2) the society's "reward system," or lack of it, (3) attitudes and concepts among the labor force, and the philosophy of the people involved in contributing their labor to production and innovation, (4) the political forces which influence the overall economy, and (5) the standards, if any, which are based on product rather than performance expectations. (Establishment of standards for products is a bureaucratic attempt to replace performance criteria with quantitative measures.)

In the United States and other developed nations, innovation generally has been mission- instead of sector-oriented, and the mission approach knocks down many of these barriers. The moon mission demonstrated what can be done when the president of the United States sets a national objective: as effort and resources are harnessed to support the mission, many traditional patterns and procedures are altered to suit. More recently, the public and the administration have recognized the critical need for the mission type of innovation in sectors where health, pollution, poverty, and educational standards are not acceptable to the American populace. Over the next few years social and technological innovations will undoubtedly be created from this new emphasis on public-sector orientation, provided adequate funds are made available. Just how private venture capital can be attracted to these socially responsive sectors is a challenge under study in government, in business, and in many U.S. universities.

If barriers to innovation are removed, then such anti-innovational influences as the traditional structure of industry or trade and fixed educational systems can be changed. A reward and punishment system can be provided which encourages innovation. Labor union practices and standards for products can be changed, and more reliance can be placed upon performance criteria. Such a setting allows the innovative brew to simmer until it can give rise to viable projects in a developing nation.

The key factor in fashioning an innovative setting is the identification and support of intellectually creative persons. Creativity involves developing, proposing, and implementing new and better solutions to problems. It is distinct from productivity, which requires the efficient application of current knowledge and existing solutions. The entrepreneur-innovator is primarily concerned with creativity, the institutional manager with productivity.

Studies have shown that creative people, regardless of the state of economic development, possess at least four intellectual characteristics: conceptual fluency, or ability to generate a large number of ideas rapidly; conceptual flexibility, or ability to discard one frame of reference for another and so change their approach spontaneously; originality, or ability to give unusual interpretations or responses to situations; and a preference for complexity. According to "The Creative Organization," a seminar report by Gary A. Steiner, the entire chain from the conception of an idea to the point of social or economic impact is not necessarily made up of creative links; however, at least one of the links usually involves a creative contribution.[1] The creative step is normally the inventive step, the initial idea, a flash of genius, or a nucleating event which triggers activities that, with proper development, cause an innovation having some social or economic impact.

While successful innovation is often attributed mainly to this first creative breakthrough, it may depend more on such considerations as marketing information, distribution opportunities, or viability of the scale-up. Most U.S. patents, for example, have not proved to be good investments or even good speculations for the inventors. It has been the scale-up and the subsequent innovative inputs that have made the original creation worthwhile economically or socially. Thus the ingredients of an innovative brew are only the starting point for a successful innovation.

Takeoff from an Institutional Base or Within a Stable System

Going systems are not only stable but also likely to resist perturbation and to counteract any influences affecting their existence. In November 1970 Eurocan, Ltd., the German subsidiary of an American packaging firm, introduced a four-day workweek in its plant outside Munich. The plant had a workforce of about 380 people and an annual business of about $4 million (U.S.). While the four-day workweek had been talked and written about in the United States for some time, it was something new for German industry. Everyone at Eurocan now works from Monday through Thursday on a 37½-hour week, but is paid at the rate of 40 hours. There is occasional overtime on Fridays. The cigarette industry, which has 16,000 workers, is the only other German industrial employer with a 37½-hour workweek, but that is spread over five days.

The interesting result of Eurocan's introduction of this American work-pattern innovation was the largely negative reaction to it in German business and labor circles. The Federal Association of Employers stated: "Economically, it is of the greatest danger . . . very critical from the point of view of industrial therapy." The National Trade Union Congress dismissed the Eurocan innovation as an isolated case and turned to problems it considered more important — longer vacations, for example. However, the weekly journal *Die Zeit* did comment favorably: "The main arguments for the four-day week, where the number of hours remains almost constant and the pay is the same, are increased productivity, less absenteeism, greater enthusiasm and less job insecurity." But this cheerful view was an isolated one; the main reactions of gloom and indifference to the workweek experiment were typical of the entreprenertia with which the old order meets a threat to the established way of doing things. It constitutes the formidable barrier which a stable system raises against the innovative start-up.

The problems that a large institution has in introducing innovations are exemplified by the history of Du Pont. This company is often named as the leading practitioner of basic and applied research. Its success as an innovator is cited as proof of the doctrine that bigness is a prerequisite to inventive capacity and success.

For nearly 100 years before 1920, Du Pont was a manufacturer of explosives and related products. From 1920 to 1950, however, it moved into other fields and took great strides in becoming a diversified chemical firm. Management conviction, commitment, and skill accomplished the changes, together with a willingness to accept innovation originating outside the establishment. In October 1950 *Fortune* rated

Du Pont's 25 most important product and process innovations for the past 30 years. Among them were viscose, rayon, Duco, tetraethyl lead, cellophane, Dulux finishes, Freon, neoprene, Lucite, nylon, PVA, Teflon, Orlon, Dacron, and titanium. The 25 innovations accounted for about 45 percent of Du Pont's sales in 1948. However, only 10 were based on the innovations of Du Pont scientists and engineers. Eighteen of the 25 were completely new products: Du Pont discovered 5, and shared the discovery of another. Seven were product and process improvements, and Du Pont was responsible for 5 of these. It would seem that Du Pont has been more successful in making product and process improvements than in discovering new products. Nylon, Orlon, and neoprene, Du Pont's major product innovations, have been largely based on technology acquired from others.

These findings do not support the generalization that Du Pont's size creates a perfect environment for inventive activity.[2] Although the company expanded research expenditures from $1 million in 1920 to $38 million in 1950, there was no proportional acceleration in important inventions. This raises a fundamental question: To what extent can a nation rely upon fundamental research by a relatively few large industrial firms? Such companies may not have adequate economic incentives for doing the work, and they may not be able to create the ideal environment for discovery and subsequent innovation. Like almost all large institutions, Du Pont has difficulty in nurturing entrepreneurial ventures, although its management has recognized this and has taken steps to change its style and philosophy toward innovation and research. Further organizational experimentation is underway to improve the record.

Large institutions need to learn not only to invest in new ventures more thoughtfully, but to refrain from imposing the parent strategy, philosophy, and techniques upon them. Investing outside the framework of the existing organization should be a very different matter from investing in an internal project. The chances of success are greater if the outside investment is made in a nonrelated area, since the temptation to impose practices and criteria on the fledgling is less. If established institutions do not learn how to create a climate favorable to entrepreneurs, they will lose their innovative people, who will spin off and go elsewhere to obtain the necessary freedom. The spinoff factor was a major cause of the Route 128 phenomenon.

Original ideas are by no means confined to companies of governments with enormous research programs. In some areas, of course, there is a critical minimum size for an effective research and development program. Nevertheless, a great number of valuable discoveries are made outside large institutions. In 1966, for example, 35 percent of patents granted in the United States went to large companies, 46 to small and medium-sized companies, 25 to individuals, and 5 to nonprofit institutions. In addition, large companies that possess patents which they find inconvenient to exploit themselves often aid in launching new companies for this purpose. In some cases the big company has buy-back privileges when the new firm has been developed into a venture.

If it is assumed that a potential innovation is technically feasible and that a market exists, the entrepreneur is faced with two major sets of problems — financial and management. In Part One the emphasis was on management problems. However, careful funding is also necessary to bring the enterprise to the public offering stage, just as management abilities are required to develop and successfully run it. Funds are needed in the forms of start-up capital, capital to help the company during its first growth period, and finally expansion capital. Requirements for financing vary widely in the less developed areas. Depending on whether the enterprise is a free-standing entrepreneurial effort, a cadre within a larger system, or a spin-off from a big institution, the array of management talents and the size of the financial support vary considerably. The role of venture capital in all these situations is discussed in Chapter 9, with emphasis on start-ups in the developing countries.

Takeoff in a Less Developed Environment

Not until the 1960s did most developing countries fully realize that scientific and technological research is such a powerful tool in the process of economic development. Within the next few years a major increase is expected in R&D activities in developing countries. Reflecting this trend, the United Nations plans to put into effect a World Plan of Action during the 1970s for application of science and technology to international development.

In most developing countries, practically all current R&D is supported by the state, with or without aid from the developed nations. According to an OAS report, for example, the proportion of all Latin American R&D that is government-financed amounts to 96 percent, with 98 percent in Argentina, 98 percent in Brazil, 94 percent in Colombia, 93 percent in Peru, and so on.[3] Most of the laboratories belong to the public sector and are under direct administration of various state bodies.

An analysis made by Jorge A. Sabato shows that in the early phases of publicly financed research, funds are ample and people are dedicated.[4] The government is usually enthusiastic about the project, and there is no shortage of staff or high-level attention. A second phase begins five to ten years later, when the glamour has worn off; even though some research results have begun to flow, second thoughts are flourishing. Staff members who have been sent abroad to survey the situation outside the country have learned a new style of endeavor. When they return, the effort at home is in an operational phase, and budgeted costs rather than new investments in facilities have become the political focus. In the ensuing conflict the government often loses its enthusiasm for R&D, especially in view of other forces that are usually at work against the relatively unstable system of a developing country. The third phase is a period of what Sabato calls "degeneration through administrative inertia." The project has accumulated a mass of mediocrity, bureaucratic red tape, and methods for suppression of any creative endeavor. The country's whole scientific and technological structure is thus seriously affected.

The delicate, unbalanced socioeconomic structure of developing countries, which must face deficits, trade balance problems, indebtedness, troubles caused by a weak currency, political upheavals, and unrest from social inequalities, makes it difficult for innovative activities or even straight development programs to survive. Ultimately, the country's economic dependence sets considerable limitations on its ability to promote continuing national research and development efforts. As a result, its R&D investment is usually in economic sectors which are directly under state control — those which contribute toward development of the economic infrastructure of communications, telecommunications, power, fuel, transport, and so forth. The tendency is for the amount of investment in these sectors to be entirely out of proportion to the budgets of enterprises which could really make use of research and development. The government often gives private economic sectors no incentives which would encourage them, such as tax relief, grants, or research and development contracts. As for firms which are subsidiaries of foreign companies, they usually do not sponsor R&D in the developing country but receive their know-how from their main offices abroad. Experience has shown that denationalization of enterprises leads to an improvement in the technology of a developing country owing to the importation of know-how, but it also increases the country's technological dependency.

A developing nation is by definition one that is in transition to a modern society and, as such, lacks an established scientific tradition, an economic infrastructure, and basic economic stability. Hence the opportunity for innovative technological activities in developing regions is limited. Development is an integrated process involving the accumulation of intellectual and monetary capital and the creation of an environment and a culture which promote the efficient use of this capital for long-term objectives. Innovators and investors cannot reasonably expect a developing society to have an order and stability that are ideal for an innovative start-up and a new venture's growth. On the other hand, governments of developing nations must recognize that scientific and technological research is becoming a new function of the state, and that it cannot be carried out within the framework, rules, and procedures laid down by nineteenth-century laws. As Sabato justly observes, most developing nations need to overhaul their legal, social, and political structures in order to minimize antiquated barriers to innovation.

Recently the industrialized nations have made a great effort to broaden the economic base of developing countries. While a large portion of this attempt has been viewed as a moral obligation that should be fulfilled by government agencies with public funds, private capital investment has played an ever increasing role. Private investment has proved an especially effective form of development aid, because it combines capital infusion with commercial and technical experience. For the developing country, development capital investment often leads to the improvement of the economic infrastructure. This induces further investment by foreign and domestic enterprises — investment which in turn reduces the dependence on imports, promotes exports, and generally contributes to the relief of balance-of-payments problems.

Incentives to the foreign investor include the opportunity to stabilize markets, utilize less expensive labor, make cost savings on raw materials or fuels, take advantage of government incentives or beneficial taxation, and secure raw materials sources. Disincentives consist mainly of political risks, such as expropriation. The wave of expropriations in Central and Latin America have concentrated on industries concerned with natural resources, while in countries like Tanzania and Uganda the focal points have been the industrial and manufacturing sector, banks, insurance companies, and transport entities.

Besides expropriation in the classical sense of deprivation of proprietary rights, barriers to innovative activity are raised by more subtle forms of government intervention in some of the developing countries. One quasi-expropriatory measure is the temporary, uncompensated requisition of the investor's property. Another is the imposition of requirements to sell products at unrealistically low prices fixed by the state. Then there are trade restrictions limiting the import of machinery or spare parts, or forcing the investor to fill his needs from a possibly insufficient domestic supply that is of inferior quality. In many developing countries the courts do not function well, which makes it difficult to predict whether foreign investment will have sufficient protection. In some nations these barriers are so formidable that true venture activity will have to wait for a long time.

In considering the investment climate in a developing country, another important issue is the employment of foreign personnel. At least during the start-up of a new project, the major portion of the staff must usually be recruited from outside. And in the case of disputes, the question of judicial enforcement of rights should be dealt with. Agreements with partners in developing countries that give sole jurisdiction to the courts of the investor's country can lead to a Pyrrhic victory: a decision in favor of the investor has no benefit if he cannot execute it.

Differences in attitude and philosophy between managers in developed and less developed areas can also affect the climate for innovation. In the more sophisticated industrialized environment, managers usually accept the fact that change is inevitable. In addition, they are accustomed to readily available information on public companies and to an interchange of business and management know-how. Sophisticated and experienced managers recognize that their enterprise is free to fail, and they understand that they need to make their hierarchical structures less rigid in order to cope with an increasing number of outside forces and to encourage innovative efforts. The more developed managerial climate puts a high value on ideas and less value on the credentials of the person presenting the ideas.

In contrast, managers in less developed areas have a tendency to maintain the status quo and to be secretive about their activities. Frequently the prevailing attitude is that the institution is immortal because it will be supported by the government, and that it therefore does not have freedom to fail. In other words, the establishment will go on regardless of the need for it. (Although this notion is unfortunately often found in more developed nations as well, it is not usually predominant.) The manager in the less sophisticated industrial environment tends to put more emphasis on his institution's hierarchy. Also in this milieu, a person's credentials often assume undue importance: he is judged by his education, status, or family background rather than by his ideas.

A recent survey of a group of international managers in both developing and developed countries shows some common denominators which affect the climate for innovation. The question concerned the most pressing problems which managers could foresee for the next ten years, and which they had not yet been able to solve. Of the more than 200 answers, only about one-third proved to be new problems to the individuals. These, ranked in terms of frequency, were as follows:

> Management development and succession
> Government relations
> Internal management problems
> Product development
> Development of a common purpose inside the firm
> Management information systems, with the computer
> Management education
> Capital structure and finance
> Long-range planning
> Training entrepreneurs as well as managers

From the standpoint of international development and the start-up of innovative endeavors, this is a good list of key problems that generally arise after the initiating ingredients for innovation—the man, the idea, and the environment—are interacting.

Whenever innovation is both deliberate and associated with the gradient of social change, it rests on presumptive evidence and the use of the scientific method. But when innovation is incongruous, either it must struggle to establish itself by a process of give and take with the environment, or it must be imposed by force, as when the victor in a revolution or a war imposes upon the defeated a new way of doing or thinking. In imposing innovative efforts on an environment or system, it must be recognized that innovations are mostly resisted out of fear and self-interest. That which is new is almost always considered to be unreasonable. Rationality has been called a sentiment in which the feeling of familiarity is joined to a sense of congruence with our fundamental hopes and desires.

9
Venture Capital Movement

I was under the necessity of forming some fixed ideas . . . in order, amidst so vast a fluctuation of passions and opinions, to concenter my thoughts; to ballast my conduct; to preserve me from being blown about by every wind of fashionable doctrine. I really did not think it safe, or manly, to have fresh principles to seek upon every fresh mail which should arrive from America. *Edmund Burke* in his speech "Conciliation with America," 1770

ONE of the major distinctions between the world's rich and poor nations is the shortage in the poor ones of monetary and intellectual capital to spur economic and cultural development. Monetary capital, properly invested, generates more capital; intellectual capital similarly begets more advanced technology and cultural development. The world capital gap between developed and developing countries is the focus of government aid programs, private investment, and intensive research by educational institutions. Despite all this effort, continuing trade imbalances, together with regional and global political struggles among the world's major powers and the Third World countries, are perpetuating the disparity in the availability of intellectual and monetary capital.

Within the developed world there are also wide differences between industrialized nations in social, political, and economic policies and in their concepts of government's role in a free market society. Their approaches to venture capital investment and to development capital financing requirements run the gamut from lack of comprehension of the economic principles and realities of a free market economy to explicit "fresh principle" government provisions to encourrage venture capitalists. The discussion in this chapter is confined to the monetary venture capital movement.

Certain trends dominate recent developments in the world's monetary capital markets. Rapid technological progress creates an almost insatiable demand from private enterprise for financial support of continuous innovation and from the world's population for improved standards of living, for integration of the international economy, and for defense and social expenditures. All compete for monetary capital to finance short-, medium-, and long-term ventures.

Funds are needed for three basic uses: start-up capital, "incubation" or "greenhouse" capital, and finally, expansion capital. The sources of venture funds fall into two categories: (1) general or traditional sources which can invest in innovation, although this is not their main business, and (2) organizational sources specializing in innovation. General sources include commercial and investment banks, insurance companies, private investors, large industrial companies, and friends of the entrepreneur. Organizations specializing in aid to innovation include venture capital firms, venture capital funds, and government-supported development organizations designed to help research and development efforts in small companies. The institutions specializing in venture capital situations were first introduced in the United States in the 1920s. It is instructive to examine this venture capital movement and see how it is different from other types of financing.

Venture Capital Defined

The venture capital movement beckons to the investor under various names: corporate fun and games, bank rolling, capital in search of an idea, seed capital, cash versus commission deals, early-round financing, capital appreciation versus ordinary income, the new management game, start-up capital, the search for genius, and ethereal gaming. Although these phrases have been used to characterize venture capital, they do not tell how to walk from here to where the pay-off is. Chris Welles, editor of the *Institutional Investor,* calls venture capital the "biggest mousetrap of the 1970s," whereas *Dun's Review* has referred to the "golden equation of venture capital." Its speculative nature attracts nouveau venturers, particularly in areas of high technology. The public recoils from being burned by venture deals, but the smart money seems to know which way to walk for the payoff.

Venture capital has been variously defined as investment in a high-risk financial venture, or investment in a start-up or new concept, or investment in a business that has no financial record of accomplishment, or investment in a turnaround situation where a distressed business presents a unique opportunity for financial rewards. Most venture capital investments are made by professional venture capitalists and wealthy individuals, although the public occasionally participates. Venture capitalists are more concerned with generating capital gains than with deriving ordinary income.

Donald H. Wheeler, an officer of the Massachusetts Mutual Life Insurance Company and active in GROCO, Inc., Massachusetts Mutual's venture capital subsidiary, defines venture capital as "the investment of capital in unseasoned companies having unusual growth potential which the investor feels will provide the opportunity for substantial gains." The dictionary definition goes somewhat beyond this; Webster says that venture capital is money invested or available for investment in stocks, especially funds invested in stock of newer, unseasoned enterprises. The expectation of repayment in profits and dividends is subject to the hazards of ownership, as distinguished from capital loaned by banks. Synonyms, according to Webster, are *risk capital* and *equity capital.*

Venture capital firms are in the business of encouraging innovations, and have a definite place in the evolution of a successful one.[1] Assuming that a potential innovation is economically viable, the innovator is faced with a financial and a management problem. The funds for research and development may be put up by him and his friends, supplied through a company's autofinancing, or furnished by a government subsidy. When the prototype stage is reached, the enterprise moves into the start-up and risky growth periods, which occupy about five to ten years. This is the time when the true venture capitalists move in. After the firm's growth has become controlled and a performance record is evident, it has graduated to the business development financing stage. Money is then supplied by companies specializing in such financing (although they are also called venture capitalists, these investors are more conservative), by commercial banks, and by investment banking firms. As growth continues and shares are issued on the stock market, financing is obtained through the public, through merger, or through long-term loans, say from insurance companies. Such loans are increasingly coupled with equity participation, either through a convertible debenture issue or through an issue with warrants attached to purchase common stock.

There are three generally recognized phases of venture capital. The first and purest version consists of personal or partnership cash investment in return for an equity in the business. The second stage usually entails private placement investment in a company having some semblance of a financial track record. The placement may be by a formal venture capital group, a mutual fund, or an insurance company, and typically involves convertible debentures and warrants or nonmarketable stock. The third phase consists of a public offering, usually of common stock but sometimes of a package that includes stock, convertible debentures, and warrants. The convertible debenture holders may have a longer investment time span in mind, and it is not always clear whether the public stock subscription is on the basis of a long-term investment perspective or a short-term speculation. Usually it is a mixture. The recent booms and busts in fast-food franchising companies and nursing home chains are evidence and results of speculative motives.

In summary, although the term *venture capital* has no precise and commonly accepted definition, conventional usage implies that it involves investment in a business enterprise where the uncertainties have yet to be reduced to risks which are subject to the rational criteria of security analysis.

The Venture Capital Industry

Speculator: One who bought stocks that went down. Investor: One who bought stocks that went up. *Malcolm S. Forbes*

As mentioned in Part One, the stimulating effect to the economy of ready venture capital for the entrepreneurs and innovators who break out in the wake of technological advances has been well recognized by the U.S. government. Its enlightened attitude has helped to create the large industry of venture capital firms that has sprung up to take the place of the wealthy individuals and families who spawned the venture capital movement.

S. M. Rubel & Company provides a venture capital service which gives subscribers financial analyses of venture capital investing and of companies receiving capital from the government's Small Business Administration. Its 1970 directory of all known organized venture capital groups in the United States contained about 400 names—more than double the number in 1969.[2] Mr. Rubel attributes this burgeoning of the venture capital industry to several trends.

First, money seeks other outlets when the stock market turns downward as it did beginning in 1969, and venturing is relatively immune to market fluctuations because most holdings are private.

Second, perhaps two-thirds of major American corporations have created new-ventures divisions, projects, departments, or subsidiaries to polarize new-venture activity and to attract entrepreneurially inclined management people. Through the subsidiary route, large companies have some chance of retaining key personnel who might otherwise spin out and start their own businesses. The subsidiaries formed by these major corporations are sometimes operated captively and sometimes at arm's length. The investment may be a true diversification, where the enterprise is unrelated to the parent company's business, or it may be congeneric to the parent's business. The management challenge differs considerably between the conglomerate and congeneric approaches.

The third trend cited by Mr. Rubel is the high profitability of successful venture capital businesses and some small business investment companies (SBIC's) in recent years. Apparently the growth of net asset value for venture capital companies greatly outstripped the Dow Jones Industrial Average of 3 percent per year during the period 1963–1969. The popularity of venture capital is understandable when records such as those of Teledyne, Memorex, and Scientific Data Systems are considered. One of the founders of Scientific Data Systems, Arthur Rock, recovered 800 times his original investment. Xerox later purchased SDS in an exchange of stock worth nearly a billion dollars at the time. American Research and Development Corporation's $61,400 investment in Digital Equipment Corporation had appreciated to $290 million on the American Exchange in mid-1970, when the market was truly battered.

Another directory of venture capital organizations in the United States was published in 1970 by Technimetric, Inc. Of the 206 firms listed, about 60 are investment bankers, 60 are small business investment companies, and 60 are private partnerships or companies that engage exclusively in venture capital activity. The rest are manufacturers or insurance companies that have venture capital subsidiaries. The average investment made by these firms is $250,000, but they go as low as $20,000 and as high as $20 million. In addition to listing the companies, the directory shows their approximate range of financing, their most recent venture capital investments, and the investment areas they prefer and avoid.

Probably the principal reason for the success of the venture capital firm is its ability to choose and move quickly on investment opportunities. John Hay Whitney & Company, which in a conservative estimate has increased its capital over twelve times in the past 20 years, has received about 8,000 applications for funds from various enterprises but only 80 have been accepted. Venture capitalists think in terms of three basic areas of risk: technical, commercial, and human. As would be expected, evaluation of the entrepreneur-innovator is the most difficult task.

The public can participate to a limited extent in this venture capital boom by purchasing shares in specialized closed-end funds and in combination funds which invest in letter stock, venture capital opportunities, and the stock of public growth companies. Certain funds, which are identified by Mr. Rubel's service, tend to be relatively inactive venture capital investors: for example, Fund of Letters, SMC Investment Corporation, Value Line Development Capital Corporation, and Inventure Capital Corporation. On occasion, however, even these may become deeply involved.

There is another group of companies which generally confine their investments to situations beyond the start-up stage, when there is less risk and, of course, less gain potential. Some of them prefer to classify their activities as business development rather than venture capital investment. Massachusetts Mutual Life Insurance Company, with its venture capital subsidiary GROCO, Inc.; Connecticut General Life Insurance Company; Donaldson, Lufkin and Jenrette, Inc.; Singer Company; W. R. Grace; Monsanto; CIT Financing; General Mills; Swift & Company; Bank of America; Wells Fargo Bank; and First National City Bank—all follow this general approach. In November 1969 a consortium of insurance companies and banks formed the nation's largest private venture capital company: Heizer Corporation, with $81 million backing, in Chicago under the leadership of Ned Heizer, the former star performer in venture capital activities at Allstate Insurance Corporation. In early 1970, New Court and Partners, Ltd., established a $50 million venture capital company with the Rothschild interests behind it. This firm is looking for small family-owned companies that have managed to survive and meet new situations despite inept management.

Back in 1958 (one year after Sputnik and a period of slowdown in the U.S. economy), Congress passed the Small Business Investment Act. At that time such investments formed a major source of capital movement, but this source has since declined in importance—owing in part to government restrictions. The Small Business Administration's provision of capital in long-term financing to small firms for expansion, modernization, and general corporate purposes is no longer the factor it once was in relation to the venture capital movement. Small business investment companies numbered over 600 in 1970, but only a few of the larger SBIC's were true venture capital firms. Less than 10 percent were publicly owned. During 1969, the SBIC industry invested $142 million in nearly 3,000 financings, and the industry showed an 18 percent return on net worth.

In addition to the buzz words used in defining venture capital, the usual jargon of the specialists in a new industry has sprung up. When these terms become more commonplace in our everyday conversation, perhaps our society will have more innovative thrust than it now possesses.

> *cheap stock*—stock at a heavily discounted price sold to the underwriter to arrange private placement or to start up investors
>
> *Chinese money*—paper instead of cash, usually a combination of debt and equity (sometimes referred to as funny money)
>
> *going down the tubes*—a new-venture company is failing

gunslinger—a high-performance, high-risk fund taking a venture capital role; the manager of such a fund

hi-tech—high technology

hot beds—environments favorable to young venture capital deals, such as Boston and San Francisco

living dead—investment situations which are neither fabulous successes nor total failures; what happens is that they seem to go on forever, and the entrepreneur has a lifetime annuity

miniprospectus—what the new-venture company is all about: the management background of its people and its financial projections

overnight multiple—the rapid effect of a public offering on the holdings of insiders

spin out—what an entrepreneur does when he breaks away from a large corporation and starts his own high-technology company

start-up situation—may mean two engineers and an MBA who want to form their own company to do their own thing (they will usually have a few *kilobucks* of their own family money invested or available)

sweetener—an equity kicker, convertibility feature, or cash fee offered as enticement to venture capitalists

But it takes more than neology to be a successful venture capitalist. The approach that seems to pay off is more psychological than statistical, more intuitive than logical. One venture capitalist claims that the job consists of calibrating people. General Georges F. Doriot, president of American Research and Development Corporation, says, "Sometimes it is even okay to go along with a Grade B idea, if the man is Grade A."

The quality of both the entrepreneur and the prospective chief executive officer is critical in a venture capital situation, and the same person is seldom ideal for both functions. The best entrepreneurs tend to be men under 35 who can bring zest and dedication to tasks where impulsive action, instinct, creativity, and innovation are more important than drill and indoctrination into rules and procedures. Brilliant innovative departures are unlikely to come from people over 35, except when these breakthroughs "are a fruition of a lifetime of ferment, as was the aggiornamento of Pope John XXIII." [3]

What distinguishes the successful entrepreneur is his genius for rewarding departure from the normal course and his unrelenting drive to create and innovate as if it were the most important matter in his life (which it usually is). Unafraid of risk, he is often oblivious of it. Venture capitalists search for men with such intense interest and talent. A few years ago, when Dr. David Leeson, an Itek scientist with an outstanding reputation, told friends that he wished to start a new company called California Microwave, Inc., the proposition was oversubscribed in 24 hours as a hi-tech issue that would be piloted by a genius. In 1970 the company had sales of several million dollars a year.

Dr. Clarence Zener of Carnegie-Mellon University, Pittsburgh, postulates that the successful entrepreneur intuitively thinks of money on a logarithmic scale.[4] As a result, the entrepreneur will invest in a venture that has a 50 percent chance of increasing his capital by a given factor and a 50 percent chance of decreasing it by the same factor. When the factor is two, for example, he stands to double his capital if the venture works out, but to lose only half if it bombs. Logarithmic thinking, according to Dr. Zener, also determines how much the entrepreneur spends on living expenses: that is, he allocates a certain fraction of his capital to them each year, regardless of how many worldly goods he has.

Whether the entrepreneur can get richer quicker by thinking logarithmically rather than arithmetically about money remains to be seen. There is no question, however, that the venture capitalist must appraise the entrepreneur's attributes, including financial acumen, as carefully as he judges the prospective chief executive and the innovative idea itself.

Venture capital is relatively new in the business scheme of things, and we can benefit by listening to the chief executive of the first public venture capital firm. General Georges F. Doriot of American Research and Development Corporation, which is now more than 20 years old, warns: "In our business, figures speak too late. When financial statements are available, irreparable damage may already have been done." [5] General Doriot says that their main mistakes at ARD have been due to

Becoming too emotionally interested in an idea or a man.

Excessive belief in apparent newness or mystery.

Competition appearing earlier than expected.

Excessive delays in foreseeing problems and/or applying corrective measures.

Pricing products too low because of the desire to "get in."

Breakup of original team or too great loyalty to original team.

Poor knowledge of cost-overhead inventories.

Business grows — head man does not.

Product and demand do not cooperate.

Technically trained manager does not replenish his black box of knowledge and is superseded by newcomer.

Greater interest in "options" than in really building a company.

Cannot wait "to go public," then watch the market, go boating, skiing, hunting, sunning and make speeches (as I am doing today instead of working at my job!)

Lack of understanding of the difference between operating profitably and having a profitably growing competitive enterprise.

Some small companies acquire too soon the aging characteristics of large ones.

Not enough foresight and drive on the part of ARD.

Few of our mistakes have been technical; most of them have been human mistakes, either in the affiliate or at ARD.

Venture Capital Pioneers

The pioneer venture capitalists in the United States were wealthy men who could risk part of their personal fortunes and who felt a moral obligation to encourage innovation for the good of the economy.[6] Bessemer Securities was established in 1942 by Henry C. Phipps, cofounder of the Carnegie steel empire. The Starwood Corporation was organized under another name in 1929 by J. Rosenwald, former chairman of Sears, Roebuck & Company. J. J. Whitney made his first venture capital investments in the 1930s and set up John Hay Whitney & Co. for this purpose after the war. The United Credit Corporation began supplying venture capital in 1932. Laurance S. Rockefeller's first venture capital operation was in 1938, and Eastman Dillon founded the Union Security Company in the same period. The investments of these early ventures proved overwhelmingly successful, and the organizations are still very much alive. Their contributions to the U.S. economy range from the development of the technicolor film process (Whitney) to the founding of Litton Industries (Rosenwald).

Undoubtedly the income tax encouraged the early venture capital movement, since there is more incentive to take risks when the government pays for a good part of the losses and taxes only a small fraction of the gains. From the vantage point of the 1970s it seems to us that in those pre-World War II days, they were "giving" away stock, bonds, land, houses, and so on. Admittedly, risks had to be taken, but the expected returns were probably greater at Depression price levels.

Prudent investments enabled these pioneer groups to multiply their capital many times over. As already mentioned, the first publicly held venture capital firm was the American Research and Development Corporation, established in 1947 by a group of prominent, public-spirited businessmen, financiers, and professors in the Boston area. A recent check showed that ARD has made about 100 investments and now retains interest in some 45 companies. ARD is also the founding shareholder in two similar corporations outside the United States: Canada Enterprise Development Corporation and European Enterprise Development Company (EED).

A Growing Free-World Phenomenon

While the basic development of venture capital as a formalized concept took place in the United States, venture activities today are a global phenomenon. Of course, the venture movement is in widely different stages in different world areas, with the industrialized countries being the most advanced.

Because of the very high risk that venture capital sources assume, an economy must have a mechanism, sanctioned by public policy, through which something is offered to the venture capitalist and to the public in return. In the United States this mechanism is the public securities market — the registration–new offerings ritual practiced through investment banking firms and administered by the Securities and Exchange Commission. This is obviously a more sophisticated mechanism than could be set up in world areas where more fundamental developments have yet to take place. Even in the more developed nations, according to the U.S. Department of Commerce,

> By and large, the technical people, who have the idea and want to build a company on it, have little if any business experience and know nothing about the venture capital market. On the other hand, the sources of capital — banks, wealthy individuals, underwriters, investment trusts, and others — usually have no technical background and only rarely have available to them adequate staffs to perform the complex investment appraisals required to measure the merit of any single entrepreneurial proposal. We are dealing here with ideas that have high technical content. The venture capitalist needs to weigh their prospects. He may have a great many new ideas presented to him. He must pick winners some of the time and make educated gambles. . . .[7]

The following descriptions of regional trends in the venture capital movement, including those in the United States, represent a sampling chosen to indicate the general status of development and venture capital financing around the world.

United States

Venture capital in the United States is affected by certain tax factors. Large companies generally have other profits against which innovation project losses can be written off immediately; therefore the government currently shares in 48 percent of these losses. But since small companies often do not make profits for five years or more, the government defers its contribution until profits are realized—or if losses persist for longer than five years, the government is not called upon to share them.

An advisory committee of private citizens was formed by the Secretary of Commerce in 1966–1967 to look at ways to remove tax disincentives and provide incentives for innovation.[8] The study made by this group found that the three main factors affecting innovation and invention are taxation, finance, and competition. The committee concluded that there was no need to recommend any major changes in present U.S. laws governing the three areas. However, it did make a number of specific proposals to improve the environment for innovation and invention. Broadly, they concern:

- Tax carry-forward of losses for small, technology-based companies.
- Liberalization of stock option rules.
- Change in IRS code to fit inventors.
- R&D expenditures for new products or processes should not be disallowed as a business deduction because they are unrelated to current activity.
- Broader studies of the innovative and entrepreneurial processes by the Department of Commerce.
- Interdepartmental ad hoc reviews of activities affecting the innovative environment by departments of the government.
- Clarification of regulatory and antitrust agency activities.
- White House conference on understanding and improving the environment for innovation.

According to this analysis, the difference between the processes of invention and innovation is the difference between the verbs *to conceive* and *to use*. Invention is to conceive, and innovation is to use—to translate an invention or idea into the economy. Innovation and economic growth are thus directly related. The need to provide incentives for innovative ventures has been widely recognized by the business community and the government only in recent years.[9]

A recent study by the Sloan School of Management at M.I.T. examined the job-creating power of venture capital and found that roughly $1,500 of venture capital investment spawned one primary job.[10] The exact amount is not important; the significant fact is that venture capital has a powerful job-creating capacity for each risk dollar utilized. Over an average period of 4.2 years, 21 companies increased their sales totals by about $77 million and their employment totals by 3,096 jobs. The initial venture capital averaged $225,000 and totaled $4.7 million. This amounted to an initial venture capital requirement of $1,525 per job.

The same study showed that entrepreneurs in science-based businesses tend to share a number of traits, one of which is the propensity for taking risks. In a group of more than 250 firms around Boston's Route 128, the investigators found that 100 companies were spinoffs from four major M.I.T. laboratories; the rest were the brainchildren of former faculty members of four M.I.T. engineering departments. The average company was four to five years old when studied, and its 1968 sales were in the $1- to $5-million range. Analysis revealed that these young companies seldom fail; the failure rate over the four- to five-year span was only 20 percent, compared with the national average of 50 percent in the first few years of a new business.

The usual statistician's explanation of why the United States is the richest, most diverse society in the world overlooks one of this country's most valuable resources: the inquiring mind and inventive spirit of the American. Some idea of the specific impact of invention can be gained by looking at two relatively recent developments: the Polaroid camera and film, and the Xerox process of dry copying. These inventions alone spawned two big companies with combined annual sales of well over $1 billion and a combined workforce exceeding 30,000 people.

A great deal of mythology surrounds inventors, and like most stereotypes, there is a grain of truth in it.[11] The world of invention does have its eccentric seekers after perpetual motion, antigravity, death rays, and the like. Serendipity, the happy faculty of accidentally stumbling across something valuable, has always played a major part in the development of new useful things. Examples are floating soap, a nonstick coating for cookware, better brakes for jetliners, a vulcanization process for rubber, and a rat poison that harms no other creature.

But a major key to the U.S. technological scene is the availability of venture capital for inventors and innovators. This is where we have led other world areas in encouraging innovative ventures. The OECD panel's overall study came to some general conclusions about factors in the total environment which seem to encourage the creation of new technological enterprises.[12] In the United States, it was found, there exist:

1. Institutions and individual venture capital sources that are at home with technologically oriented innovators and have the rare business appraisal capabilities necessary to diagnose the prospects of translating a technical idea into a profitable business.

2. Technologically oriented universities located in areas that have a business climate which encourages staff, faculty, and students to study and themselves generate technological ventures.

3. Entrepreneurs who have been influenced by examples of entrepreneurship (for it is the panel's contention that entrepreneurship breeds entrepreneurship).

4. Close frequent consultation among technical people, entrepreneurs, universities, venture capital sources, and other agencies essential to the innovative process. Viewed in this sense, unsympathetic bankers, inattentive educational institutions, overzealous tax authorities, and other environmental barriers are negative charges at work against the entrepreneur.

Latin America

In Latin America, the bulk of funds for technological and other venture projects is provided by special development banks. These include the Inter-American Development Bank, Banco Industrial in Chile, Corporación de Fomento in Brazil, Corporación Venezolano de Fomento in Venezuela, and Peru Investment Company. IBEC, the Deltec Banking Company, and ADELA have also done extensive venture work in Latin America.

ADELA (Atlantic Community Development Group for Latin America) has been called by *Business Week* "the company that nurtures Latin business," for it provides money, know-how, and management. Owing to the circumstances in this world area, ADELA has a somewhat different approach and objective from the usual venture capital organization: it is characterized by an emphasis on social and managerial innovation rather than on technological innovation. Borrowed licensed technology is generally adequate for Latin American ventures, which face other environmental problems. From 1964, when ADELA was organized, until mid-1970, it put share capital amounting to $82 million into 99 companies in 19 Latin American countries. ADELA claims that it has helped mobilize close to $1 billion worth of new venture investment, and that for every dollar invested by ADELA, at least $9 is invested by banks, lending institutions, and other companies. This generally involves long-term, low-interest credit, as a rule indirectly guaranteed by a Unites States government agency or quasi-government agency. Such financing is usually classified as an industrial development type, with different risk-return criteria from those which the usual venture capitalists reckon by.

Two hundred and thirty-nine of the largest banks and industrial companies in the United States, Canada, Europe, Japan, and Latin America back ADELA with technical and managerial know-how, and each has subscribed up to one-half million dollars. The study leading to the creation of ADELA was financed by the Ford Foundation. Many of the

same shareowners are helping organize the Private Investment Company for Asia, dedicated to financing new ventures
in developing nations of the Far East. ADELA's aims go
beyond venture capital per se and include multinational
integration, improvement of the business climate, formation
of smaller investment companies in Latin American countries (seven have been established so far), and the eventual
creation of mutual funds. The venture capital movement is
not yet a significant factor in Latin America, but it will become one when the region develops further economically.

Western Europe

It is still far more difficult for companies and entrepreneurs to get financing started on the European side of the
Atlantic than in the United States. Up until recently, the slo-
gan has been — Eurobrains: who'll buy them? The situation
is bad enough in Britain, but on the continent it has been
characterized by *The Economist* as nearly hopeless.

Continental Europe. An OECD study focused on the electronic components industry, a supreme example of a technologically based business needing venture capital support,
in an attempt to find out what was wrong with electronic
components companies outside the United States.[13] Companies have to stay in the top rank of industry to survive;
once their expertise falls behind the field, they find themselves out of business. Even in a company as powerful as
Philco, which in 1955 was one of the top three American
firms, profits dropped by 90 percent in the following years,
and in 1961 Philco was taken over by Ford Motor Company.
The history of the industry in America is also marked by an
unusual number of cases in which companies have appeared from nowhere and skyrocketed to the top.

Of the big 10 American component makers 13 years ago,
when the transistor revolution was at its peak, only the number one company, Bell Telephone, occupied the same position eight years later. The rest had all moved down the list;
half were no longer there at all. The new number two com-

pany had previously been ranked seventeenth; the new number three had previously been twenty-second. The industry's general instability appears to have two causes: one is the complacency of some of the established American companies, which leave gaps for alert young scientists to exploit. Such complacency is by no means confined to American companies, but the second cause is unique to the United States. This is the ease with which three or four scientists working together in a laboratory (usually in an established company, sometimes in a university) can decide to throw up their jobs, set up a business of their own, and without serious problems, raise the finances to do it through the many readily available sources of venture capital in the United States. The availability of government R&D contracts for many new ventures is a key factor also. This favorable climate compounds the American entrepreneur's drive and opportunity.

Time and again, according to the OECD report, the supremacy of American industry in science-based technology can be traced to the fact that it is much easier to start such a small company in the United States than it is in Europe. And not only to start it, but to get help for the equally crucial second stage of growing and expanding production before the competition can catch up.

In addition, one of the biggest single influences on electronics technology has been the presence in the United States of a huge domestic television market. Color TV demand has provided some of the strongest drives for improved component design. Thus the buoyant, coherent consumer market's high standard of living and its taste for exotic consumer goods are two of the prerequisites for a thriving electronics industry. This does not necessarily have to be the home market, although it helps. But technology can flourish outside the American climate, as is shown by the growth of the Japanese SONY enterprise on very much the American pattern. SONY started after the war with a staff of 36, a capital of only $500, and a belief in spending heavily on research. It now has over 10,000 employees and sales over $300 million, 60 percent of which comes from exports.

One of the pioneer venture capital corporations in Europe is the European Enterprise Development Company (EED), whose founding shareowner was the American Research and Development Corporation (mentioned earlier as the first publicly held venture capital firm). EED is perhaps the most internally oriented of the venture capital groups in Europe. It was founded in 1965 by 50 financial organizations in 18 countries, and 70 percent of the capital was subscribed by 43 European banks. By early 1970 EED had invested in 25 companies in 9 European nations, with only one investment a failure.

Sweden Incentive AB, an arm of the Wallenberg empire, is somewhat similar to ARD. Britain, of course, has the National Research Development Corporation (NRDC) which was set up by Parliament. NRDC's twenty-first annual report, issued September 1970, listed 481 projects, with another 84 authorized but not yet financed. NRDC had combed through 22,960 proposals to find the 565 it accepted: a 2½ percent yield over the years taken to reach its "majority." An organization similar to NRDC, called ANVAR, is now functioning in France.

Institutions like ANVAR and NRDC differ from the corporations such as Eurofinance, which is headquartered in Paris. Started in 1961, this company serves the need for a Europe-based, Europe-oriented, Europe-experienced "partner in planning." The planning is particularly financial, but it also includes investment advice research and economic and industry research. While Eurofinance features assistance to multinational companies, it also analyzes venture companies and makes financing arrangements. This type of European organization is not classified as a primary venture capital firm but as an enterprise close to the traditional banking community's activities.

United Kingdom. Like those in Continental Europe, special institutions providing venture capital in the United Kingdom are of fairly recent vintage. In the United Kingdom, funds for such purposes have traditionally been furnished by the merchant banks; on the Continent they have been supplied by private banks and, especially in Germany and Italy, by large deposit-taking institutions.

The provision of equity and loan capital for small firms has always formed part of merchant bank business in the United Kingdom. In the early thirties, the MacMillan Committee concluded that there was a gap regarding facilities for small and medium enterprises. As a result the Charterhouse Development Company was created at that time, and after the war two new institutions, Finance Corporation for Industry (FCI) and Industrial and Commercial Finance Corporation (ICFC), were established for the same purpose of giving financial support to small firms.

The emphasis on providing venture capital for technological innovations is fairly new. Apart from the Technical Development Corporation (TDC), set up originally as a private venture but subsequently taken over as a subsidiary by ICFC, this type of business is still largely handled by the merchant banks. (Belgium's Societé Nationale d'Investissement is a similar institution, set up to nurture new ventures until they can be marketed.) Four years ago, three leading United Kingdom trust groups combined to form Associated Trust Holdings (ATH), whose major objectives are to deal with the problems of investment in unquoted companies.

Another new company called Scientific Enterprise Associates was also formed recently in the United Kingdom. Ronald Grierson, who was at one time head of Britain's Industrial Reorganization Corporation, set up the organization with Italian, German, and French sponsors and with financial backing from S. G. Warburg, N. M. Rothschild, and Societé Générale de Belgique. Roger Brooke is managing director, and the objective is to finance organized groups and manage small companies working on advanced technologies. The motivation appears to be the opportunity for substantial profits rather than a sense of public duty.

In a recent U.K. report, technological innovation is defined as the technical, industrial, and commercial steps leading to the marketing of new products and to the commercial use of new technical processes and equipment.[14] The availability of ready venture capital to support innovation was the second most important of four distinct differences named between the U.S. and U.K. environments for innovation. At one extreme in the United Kingdom, innovation implies simple investment in new manufacturing equipment or any technical measures to improve methods of production. At the other extreme, it means a sequence of scientific and market research, invention, development, design, tooling, production, and marketing of the new product. According to the U.K. view, a major purpose of technological innovation is the commercial exploitation of technical knowledge in order to win new markets or to hold existing ones against competition and to reduce production costs. Scientific advances generate the invention, which provides the "technological nucleus of the innovation," which in turn supplies the opening for venture capitalists. The problem as seen by this British study group concerned the American lead over the past two decades, during which firms based in the United States originated two-thirds of the successful innovations and produced about one-third of the world's exports of technically advanced goods—the largest individual share among all countries. Five conditions for successful innovation emerged from this U.K. study:

(a) The direct linkage of the research and development activities to the financial and marketing activities of the organisation as a whole;
(b) framing of planned programmes of innovation in relation to the assessment of opportunities revealed by a sophisticated analysis of market situations;
(c) management which not only is effective technically, but which is market-oriented and dedicated commercially;
(d) a capability for achieving short lead-time from the start of a new project to the marketing of the initial product; and
(e) a proper scale of production capacity and size of market in relation to the launching costs of the project.

The idea that R&D, production, and marketing should come under the same control so that they constitute a single innovative activity has not been appreciated in countries of Western Europe as much as in the United States. A recent analysis of major industrial innovations shows that of the inventions which led to them, 10 were initiated by Britain, France, and Germany and 19 by the United States. Only 7 were converted into final product innovations by the European countries, as against 22 by the United States.[15]

According to the U.K. report, American industry has four major strengths. First is its ability to carry an idea through to the final product without a break in the innovative chain. Second is the readiness of banks and private investors to finance technological innovation. Third, the scale of the U.S. government's purchasing policy since World War II, with its huge military establishment which has constantly demanded more sophisticated equipment, and has been an extremely important force in raising domestic demand in general and in generating the sense of technological awareness which characterizes America's vast home market. Fourth, effective leadership by management and technical leadership from management creates enthusiasm and determination in forcing inventions through to success.

The U.K. study group does not mention a fifth strength, which stems from America's twenty million or so public investors with their fantastic information disclosure and reporting network, the equally fantastic savings and dollars per capita ratios, and the American appetite for high risk–high reward stocks.

There are a number of significant differences between capital markets in the United States and in the United Kingdom, particularly in respect to venture capital opportunities. Venture capital companies have been relatively slow to develop in the United Kingdom. Donald H. Korn of Arthur D. Little, Inc., points out that venture capital financing as known in the U.S. derives its attraction from two factors: the characteristics of the risk-return curve as perceived by (and in recent times, realized by) the venture capitalist-entrepreneur, and the coupling of venture financing to the public securities markets, which allows capitalization of portfolio values through public offerings at some appropriate early stage in the venture's life—all with relative ease.

The U.S. public's acceptance of new issues and enthusiasm for "owning your share of American business" generally has resulted in new offerings selling at a substantial premium over their book value as stated according to generally accepted accounting principles. The premium is related both to the high valuation placed on earnings of companies "going public," because of the growth expectations, and to supply-demand factors which typically exist when such a new issue is underwritten by investment bankers. This fact of life in the United States has two primary causes: (1) The pattern of consumer savings and the level of discretionary income is higher here than in other developed countries— certainly in an absolute sense and probably in a relative sense. (2) The Securities and Exchange Commission, the Securities Act of 1933, the Exchange Act of 1934, and the Investment Company Act of 1940 created a mechanism in the United States for routine transfer of ownership of equity securities, together with widespread disclosure and dissemination of corporate information and readily available quotes on share prices, both over the counter and on the securities exchanges. The OTC market is the key. Thus the situation in the United States is conducive to venture capital financing in the usual context—with the existence of a "new issues market" making it possible for entrepreneurs to establish large capital gains at an early date.

Venture capital, some people in the United Kingdom argue, is no more than a new name for a process as old as joint-stock development. Yet the venture capital movement is of growing importance in the United Kingdom, because three-quarters of the firms in Britain employ less than 100 people and because there have recently been rapid advances in new business techniques and services.

The U.K. situation differs from that in the United States in several ways. The tax structure makes it more difficult for outsiders to enjoy the kind of speculation in new issues that has occurred in this country. Short-term capital gains are defined in the United Kingdom as holdings of one year or less and are taxed as unearned income, so that the top tax rate is reached very quickly. (Long-term capital gains are taxed at 30 percent or the individual's tax rate, whichever is lower.) More important, of course, the consumer savings and discretionary income stream is relatively small. Consumers invest in unit trusts, which are the equivalent of our mutual funds, rather than directly in the securities of individual corporations. It is mainly wealthy private individuals who own shares directly. Furthermore, although there is a need for venture capital, the United Kingdom has no public over-the-counter market to readily capitalize the worth of the venture's securities and substantially reward the entrepreneurs and venture backers.

In the United Kingdom, going public is more or less synonymous with having shares "quoted" by members of the London Stock Exchange, which thereby involves a listing. In the United States a company that goes public is not required, either formally or informally, to have its shares listed on any exchange. The much more heterogeneous group of broker-dealers and market makers in the United States facilitates OTC trading.

According to Mr. Korn, by the time a U.K. company is ready for listing on the London Stock Exchange (technically "to be quoted" by members of the London Stock Exchange) it is a mature organization unlikely to show the dramatic increase in sales, earnings, and stock price that might be expected from a young U.S. corporation going public. In practical terms, nobody goes public unless "trading" profits are greater than or equal to £150,000 per year. A five-year "track record" is also needed except under exceptional circumstances. Apparently, then, a portion of the risk-return curve is unavailable to the outside investor in British equities — or perhaps the curve is different. The relatively large percentage gains from capital appreciation in a speculative public new-issues market would seem to be relatively infrequent under the U.K. system, which is actually more conservative than that in the United States.

At least three main categories of venture capital investment opportunities are common to the United States and the United Kingdom. (1) High technology areas: This category, generally associated with government-sponsored research and development, provides the classical background for new company formation and venture capital financing in the more exotic areas of U.S. industry. It is still a small part of venture capital activity in the United Kingdom, because the government's National Research Development Corporation functions as primary stimulus to new ventures involving government-sponsored R&D and is empowered with the mechanism for initial financing of such ventures. However, direct government participation in new ventures is likely to diminish. (2) Franchised small businesses: A small category, but in the United States a growing one. (3) Real estate: This is a much more widespread area of activity in the United Kingdom. A fourth venture capital opportunity in the United Kingdom may be the area concerned with realizing assets from companies that are not making full use of them.

In the United Kingdom as well as the United States, institutional investors and wealthy private family groups are usually offered the initial equity deals. In the United Kingdom, however, the shares are listed or quoted by the time the public can participate, and the company is too mature for the dramatic gains that would be expected from successful (or highly speculative) younger companies. Agreements and conservatism among the council of the London Stock Exchange and the council of British Stock Exchange firms apparently further reduce the opportunities for marketing speculative securities.

In the United Kingdom, according to Mr. Korn, the name of the game is not exactly venture equity investing but rather venture financing. The latter, which resembles industrial development financing in its risk-return characteristics, is significantly different in nature.

Middle East

The venture capital movement in the Middle East varies from a very active program in Israel to an evolving or non-existent one in some Arab countries. A fairly recent movement in Israel is designed to develop a "Route 128 phenomenon" by encouraging science-based industry as a major support for the economy. In 1967 Israel Research and Development Corporation, Ltd. (IRDC), was formed to foster scientific and high-technology industrial projects, to supply venture capital for companies in this field, and to participate and guide the management of such enterprises. Participants in IRDC include the government of Israel, leading Israeli banking and investment institutions, and private investors from outside Israel. New projects are evaluated in cooperation with research and academic institutions, and IRDC encourages joint relationships with foreign companies experienced in development, manufacture, and marketing in the relevant fields. IRDC finances a project by supplying equity capital, if it is needed, supplementing it by long-term loans, but the form of investment is highly flexible.

Although there is no direct relationship between IRDC and the American Research and Development Corporation in the United States, the conceptual approach in Israel is similar to ARD, and some of the participants in Israel are also active in some of the ARD projects.

Sidney Musher, an industrialist from New York who is chairman of IRDC, is an outstanding inventor himself and has given much thought to the question of whether it will become financially attractive for Israeli inventors to have their patents developed for the home market instead of abroad. In the past Israel has suffered from what might be called an idea drain in addition of the normal brain drain; that is, ideas and inventions as well as talented professionals have emigrated to better-paying countries. The escape of Israeli ideas to countries that are willing to pay for them is one reason why IRDC has been formed to supply venture capital. In 1970 IRDC had current investments in 10 active new enterprises which were widely diversified in the high-technology field.

In the Arab Middle East, there is so far no major organized effort to secure and place venture capital, although Arab governments are aware of the need for such an effort. Indeed, they are taking steps to encourage the productive placement of surplus funds of their leading entrepreneurs in ventures which are by and large initially sponsored by government feasibility studies and seed capital. Arab governments are also aware that in the past some of their wealthy citizens have provided venture capital to the more developed areas of the world.

The definition of venture capital must be adjusted in the context of the Arab Middle East, for according to Dr. Robert G. Wilson, Aramco economist, this region considers the terms *high-risk financial venture, start-up capital,* and *bankrolling* as applicable to traditional industries as well as to high-technology ones. The establishment of a steel mill, a cement plant, or a paper factory might be considered legitimate venture capital undertakings in the developing world. Financial experts may dispute this modification of the definition of venture capital, but if they do, they are doing so in terms of the economics of the developed world.

Owing to a lack of research facilities, machine tool industries, and the like, the Arab Middle East is generally incapable of articulating a new technological concept for itself at the present. (This contrasts with the situation in occupied Palestine, where skills have been imported with the immigration of recent years.) Thus efforts to raise Saudi Arabian "venture capital" are concentrated in the cement industry, the steel industry, and tertiary petroleum industries. The importation of this technology is often sponsored by government-to-government deals. Often the projects receive government support, or in the petroleum-producing areas they are frequently the outgrowth of pledges included in concession agreements. It is the government which usually employs people with the skills to audit feasibility studies performed by consultants, and which has the moral suasion to rally the capital of local entrepreneurs. It is not surprising to find, then, that Arab Middle East venture capital projects are under the aegis of government planning efforts.

The governor of the Saudi Arabian Monetary Agency in an October 1970 report to His Majesty, King Faisal, called on the citizens who own capital to invest it inside the country and increase their savings in order to contribute "side by side with the State to the success of the development plan which the State has started to implement." [16] This was a plea to Saudi Arabian capitalists who transfer their capital abroad to invest it instead in Saudi Arabia as an aid to the economic growth of their country during the next ten-year plan. The governor's report pointed out that the surreptitious transfer of capital outside not only withholds money from the country, but also implies lack of confidence in the government. According to the governor, "nowhere else has capital been given such permanent protection as it has been given in Saudi Arabia, in which the protection of capital is a cornerstone of the State's policy."

The 1969 annual report of the research department of the Saudi Arabian Monetary Agency refers to the government's evolving interest in venture capital through specialized institutions over and beyond commercial bank activity. The banks provide medium-term credit, but their activities are limited to classical banking risks and functions. In recognition of the need for venture capital, SAMA has established an industrial bank — a bank for people of small means for the promotion of growth in the industrial sector.

By 1969 the Saudi Arabian Ministry of Commerce and Industry had prepared feasibility studies of 27 new-venture projects believed to have good chances for success. This was an effort to "overcome the industrial shyness of Saudi businessmen and to encourage them to invest in suitable industries." The studies were given to local and foreign businessmen, and some ventures — for example, in PVC pipe, polystyrene, and the canning of vegetables and fruits — are under way. During the years 1964 to 1969, the ministry granted licenses for the establishment of 191 new-venture industrial establishments, of which 144 were Saudi-owned and 47 were either foreign or joint Saudi-foreign enterprises. A total of Rs 107 million investment was forecast, and these new enterprises were expected to employ around 4,500 persons.

Saudi Arabia's currency stabilization program was brought to a successful conclusion in 1959, when the currency was declared to be convertible. For centuries the vast area presently covered by Saudi Arabia, with its predominantly tribal structure and its extreme scarcity of water, lived on the margin of subsistence. The problems involved in the process of economic development have indeed been formidable. With the discovery of oil and the more meaningful direction of policy during recent years, impressive advances have been made. GNP increased almost threefold over a decade, with an overall average growth rate of about 12 percent per annum. Per capita income has risen from Rs 775 to Rs 1,850. An outstanding feature of development has been the activation of the private sector and the very early beginnings of a venture capital movement.

Private capital formation has increased about 10 percent per year and has more than doubled over the last decade. While the government has provided incentives for industrial expansion, the general policy has been to set up mixed public and private industrial enterprises. The banking system has been consolidated, and the planning machinery of the government has been reorganized, with priorities on allocation of resources. For the next ten years, it is expected that Saudi Arabia will give further encouragement to venture capitalists to invest within the region, and will try to discourage them from going outside for venture capital opportunities.

South Africa

In recent years South Africa has had buoyant economic growth in spite of many social problems and world pressures. Industrialization and expansion have been going at a rapid clip in this part of the world, and an expansionist mood should continue for some time. The economic growth has a complicated ethnic and political background involving the two major white population groups in South Africa, namely, the Anglo-Saxons and the Afrikanders. Before World War II, the Afrikanders did not play an important part in commerce and industry. After the war, however, this group developed great drive and an unorthodox approach that has resulted in its economic emancipation at a fantastic rate. Afrikanders now figure strongly in all areas of the economy. The group also wields substantial political power, having achieved this leverage in 1948.

With the diluting of the influence of the Anglo-Saxon sector, some other trends have been evident in South Africa. Until comparatively recently, capital for development or new ventures was available to the striving Afrikander only from financial and government institutions specifically established for this purpose. But from 1967 to 1970 more than 200 companies went public, and there was an insatiable demand by general investors for their equity. Thus there is no shortage of financing for any company that has good ideas. If a firm is too small for normal public offering, it has been easily and rapidly taken over by larger companies in exchange for equity.

There has been no well-publicized desire by South African financial institutions to offer financing for what in the United States is termed venture purposes. The money market is administered through, and/or utilizes the funds of, modern sophisticated commercial banks, merchant banks, insurance companies, pension funds, and a grey market. Merchant banks certainly have placed secured or unsecured notes, as the circumstances warranted, with interest rates appropriate to the borrower and sometimes with convertible rights. Where a need arises that is acknowledged to be in the national interest, a government-financed body, the Industrial Development Corporation, lends money either with or without convertible rights and exercises a watchdog interest. The evidence of a coherent capital market in the true venture rather than the economic development sense is yet to be formed in South Africa.

Asian Pacific

In the Far East, attempts are being made to create nationally based development banks (for example, in Thailand and India) as well as regional banks sponsored and supported by governments, international institutions, and private banks. In Japan, a law similar to the U.S. Small Business Investment Act was passed in 1963, and three small business investment companies have since been set up. Venture capital groups as known in the United States have yet to be established in Japan, however. The economic-political structure there is such that any venture capital movement would have to evolve in a form entirely different from that which is common to the free enterprise market economy and industry-business-government relationships of the United States.

There are many cases in Japan where a market need for raw materials, services, products, or technology has generated a coordinated response, followed by full capital funding by major zaibatsu group banks or trust companies. It is common to have two or more companies cooperate, first in R&D to search for a product or service, and then in the development financing. Successful collaboration of this type has taken place in many fields: synthetic paper, salt from Australia, coal, iron ore, and bauxite from Australia, woodchips from Alaska and Canada, timber from Indonesia, oil exploration in Alaska, Abu Dhabi, and Indonesia, offshore oil in Japan, electronic specialties, and computer centers and software development.

However, this kind of activity, with its subsequent capital funding by the participants, is usually the opposite of what is generally considered the venture capital approach. The Japanese style of collaboration involves a need or an idea looking for capital rather than the venture capital mode of capital looking for an idea. The traditional Japanese approach sets up the new venture with a limited capital base, and the rest of the money is obtained by borrowing from banks. Even when money is tight, investment funds needed to support current national objectives are obtained fairly easily.

An example of classical financing for developing projects in less developed world areas is the activity of the China Development Corporation (CDC), the chief agency in the promotion of "venture capital" projects in Taiwan. The company is about 70 percent privately owned and 30 percent owned by the banks in Taiwan. Other than the "limited branches" of FNCB and the Bank of America, these banks are government-controlled. The original funds came from the International Bank for Reconstruction and Development (IBRD). Because of the limited industrialization in Taiwan at the time of the creation of CDC, the corporation was geared to provide development capital rather than true venture capital as it is characterized in the United States. CDC investment was and has been in economic development situations with foreign participation, and in projects where dividend or interest income is relatively certain.

The size of Taiwan (which has a population of barely 14 million and a per capita income of about $258 U.S.), and the fact that the major part of all industry is geared to export, prompts the major requirement for development capital that is channeled into known industries in known markets. CDC has made loans to the electrical, textile, paper, machinery, and fibers industries. Taiwan's first steps were taken with borrowed technology from Japan, Europe, and the United States, and industrial research per se is yet to be generated at home.

By early 1970, CDC had received a total of 880 applications, of which 374 projects involving 415 loans had been approved in its 11 years of existence. It had about $40 million out in loans, with an equity position of about $5 million in 20 different companies.

Down Under, the Australian venture capital movement was nonexistent, or essentially so, until late 1969. Today there is at least one group operating for the purpose of seeking out, promoting, financing, and consulting on ventures concerned with "scientific or technological innovation in any field of activity." The Australian Innovation Corporation (AIC) has 30 shareholders consisting of various financial institutions and industrial and mining organizations in Australia. The idea for AIC came from Dr. H. C. Coombs, governor of the Reserve Bank of Australia. In his Shann Memorial Lecture in 1963, Dr. Coombs highlighted the growing need for an Australian enterprise capable of promoting the commercialization and industrialization of local research and development efforts. A group representing interested organizations met in 1966 and worked toward the formation of AIC, which was established in October 1969 with an authorized capitalization of $1 million. Staff was assembled and an office opened July 1, 1970, in Melbourne.

The economic development of India has been substantially assisted by foreign-government aid programs rather than by capital from private sources or true venture capital. U.S. assistance to India goes back at least to 1950–1952 through the Export-Import Bank and the Bank Development Loan Fund, subsequently called Agency for International Development (AID). The World Bank and German and U.K. overseas development and loan activities are representative major donors, who have assisted particularly in the development of Indian agriculture. Venture capital in the context of more highly developed industrial nations is not yet an important factor.

India's fourth five-year plan includes aid programs which constitute 17 percent of the total outlay of the plan. The objective is to reduce foreign aid by half at the end of the current plan. The Peterson Committee, a task force reporting on U.S. foreign assistance for the 1970s, suggests the setting up of a development bank to channel AID funds and to professionalize the AID giving function in India and other developing and emerging nations.

The Indian government is awakening to the need for encouraging venture capital, not only from outside government aid programs but also from private sources. The Inventions

Promotion Board, a part of the Ministry of Industrial Development and Company Affairs in New Delhi, is one of the organizations which have recently been created to encourage this.

According to the Indian Investment Centre in New Delhi, banks and financial institutions are willing to furnish up to 100 percent of the project requirements of technical entrepreneurs who want to use their know-how to start industries. The assistance may go as high as 3 lakhs of rupees, depending on the constitution of the firm, and is in the form of a loan or of both a loan and the underwriting of share capital. In evaluating these projects, expert committees judge the know-how and the ability of the entrepreneurs. Even in medium-scale industries (that is, industries up to a capital of 1 crore of rupees), technical entrepreneurs are encouraged to enter, and their know-how is sometimes given recognition by the issue of suitable shares. This has been done in one or two cases in which institutions like the Industrial Credit and Development Corporation of India Ltd. and the Investment Bank of India have participated. When investing in such projects, the banks and the institutions do not go by normal banking considerations; they are prepared to take the necessary risks.

Seven nationalized banks have special schemes for financing technical entrepreneurs: the Central Bank of India, Bank of India, Punjab National Bank, Bank of Baroda, United Bank of India, Allahabad Bank, and Bank of Maharashtra. The State Bank of India was the first to start an entrepreneur scheme for "financial craftsmen and qualified entrepreneurs," under which assistance is provided for the entire range of financial requirements and can extend up to 100 percent. The ceiling is 2 lakhs of rupees for a technical entrepreneur who is setting up a project and 3 lakhs for two or more technical entrepreneurs who have joined to establish a project. The seven nationalized banks have similar schemes: The Central Bank of India's ceiling is 2 lakhs, the Bank of India's is 1 lakh, the Punjab National Bank's is 2 lakhs for an individual and 3 lakhs for more than one person, the Bank of Baroda's is 2 lakhs for engineers with experience and Rs. 25 thousand for inexperienced ones, the United Bank of India's is Rs. 75 thousand, and the Bank of Maharashtra's is Rs. 25 thousand for working capital and Rs. 50 thousand for a medium-term loan. In all these schemes, the entire financial requirements of the technical entrepreneurs can be met by the banks. The assets acquired are taken as security, and collateral security or guarantees are desired but not insisted upon.

Another Indian government program, this one under the Ministry of Education, is the National Research Development Corporation of India. This organization was established to secure maximum utilization of inventions resulting from the national research effort. NRDC, which is based in New Delhi, serves as a connecting link between research and industry. It hopes to stimulate development of patents and inventions arising from research conducted with public funds and, where feasible and in the public interest, to encourage the commercialization of the patented inventions of individuals also. The corporation has no statutory powers to acquire industrial property rights; it does so only by agreement and negotiation. It arranges large-scale trials and the cooperation of industry, sponsors and finances pilot-plant investigations, and performs a licensing service for industrialists. Pilot-plant work that the organization has already financed and sponsored include projects in major industries, such as textiles, cement, sugar, chemicals, and paper. The corporation has reciprocal arrangements with counterparts in the United States, United Kingdom, Canada, and Holland.

In 1968, 971 inventions had been considered by the National Research Development Corporation, of which 120 had been released free to Indian industries, 223 had been licensed to various parties, and 124 are producing under these licenses in India.

Finish Line for Venture Capitalists

Forbes (June 15, 1970) calls the stock market the finish line for venture capitalists, for those who bankroll new enterprises depend in the end on the stock market to take them out at a profit. It is assumed that once a company is on its feet, stock will be sold to the public. Usually such an event is at least three to five years away from the seed-money stage. The venture capital for this early-round financing precedes the kind of long-range money that is invested when a stock market payoff is not expected for six to ten years or more.

Although venture capital cannot be defined by virtue of whether stock is publicly available, the term commonly refers to private investment in the very early stages of an innovative enterprise before there is a record of performance. Accordingly, a venture capital situation requires greater input of management time than a normal investment, which is competitive, performance-oriented, and has an established track record.

When Mrs. Joan Whitney Payson bought into the New York Mets in 1961, the team certainly lacked a good record, and most people thought that her objective was to establish a tax loss. Mrs. Payson, a founder in 1946 of the pioneering venture capital firm Payson & Trask (Minute Maid, Intercontinental Systems, and Kingsberry Homes were among its projects), invested $3 million in the Mets for the 1962 season, and the Mets went on to win the World Series and paid off handsomely. *Dun's Review* rates the Mets as an eminently successful capital venture and quotes the reason Mrs. Payson gives for its success: "Involvement—that is, close involvement to make the investment work."

Venture capitalists like Frank Chambers of Continental Capital Corporation use an approach called "performance settlement," which is different from the competitive performance-oriented investment. Under Chambers' scheme the entrepreneur's share of the stock depends on the performance of the enterprise and acts as a kicker: the better the venture does, the more of the company the entrepreneur-innovator receives in the form of ownership.

The time of the venture capitalist's entry into a situation dictates his expectation in terms of capital appreciation. If he puts up his money at a very early stage, he may well expect a minimum 10-to-1 payoff, over a three- to five-year period, whereas if he comes in after a company has struggled for a few years, he is probably seeking a 3-to-5 appreciation payoff potential. Underlying those expectations is a reliance on the historical relations prevailing in the public securities market, including the public's appetite for equity participation in America's exciting new businesses.[17]

Some venture capitalists insist on opportunity for financial control of the enterprise in case of trouble, so that they do not have to argue money questions with an entrepreneur-innovator whose knowledge of finance may be the least developed side of his talents. Venture capitalists disagree as to the stage at which such assistance is appropriate. They agree, however, that management structures may differ as the company grows. With maturity there is almost invariably the evolution of a more formal management structure with tight financial controls, either voluntary or imposed. One of the frequent worries of bankers and some venture capitalists is the lack of such controls in some of the companies they are involved with. Just as it is inappropriate to tolerate an absence of controls in a mature company, their presence is often inappropriate and may stifle a young entrepreneurial company. Lack of understanding of this point can cause the venture capitalist to miss an opportunity.

Venture capitalists, of course, face losses as well as gains. A review of a number of venture capital firms showed that a range of 25 to 40 percent of their investments went sour.[18] In addition, the study revealed that an average of six or seven years was needed for venture capital businesses to mature their investments to a point at which they can be sold at a profit. According to the report, the minimum viable size of new venture capital firms was about $1 million, probably with more than $2 million in investable funds to support a staff and financial reserves. Many of the larger venture capital institutions tend to make investments in the range of $250,000 to $500,000, which would raise the minimum size of a portfolio to about $2.5 million.

We opened this discussion with Edmund Burke's concern

over being blown about by every wind of fashionable doctrine and his apprehension about fresh principles arriving from America. The fresh principles of the venture capitalists, despite their origin in the United States, are gaining acceptance throughout the world as a vital force in the innovative process.

REFERENCES

Chapter 1

1. American Association for the Advancement of Science Symposium, Boston, December 29, 1969.
2. Speech before the American Bar Association, Dallas, August 1969.
3. Donald A. Schon, *Technology and Change* (New York: Delacorte Press, 1967).
4. William James, "The Power and Limitations of Science," *The Philosophy of William James* (New York: Modern Library, 1925), pp. 197–198.
5. Schon, op. cit., p. 41.
6. Richard Karp, "Education: A Show of Power Over Funds for Innovation," *Science*, March 27, 1970.
7. B. Lamar Johnson, *Islands of Innovation Expanding: Changes in the Community College* (Beverly Hills, Calif.: Glencoe Press, 1969).
8. *Business Week*, June 20, 1970, p. 85.
9. *The Survey of an Emerging Service Industry: Technology Transfer*, TTA Information Services Company (San Mateo, Calif., 1970).
10. London *Economist*, August 1, 1970.

Chapter 2

1. Stafford Beers, *Management Science* (New York: Doubleday, 1967), p. 27.
2. Claude F. George, Jr., *History of Management Thought* (Englewood Cliffs, N.J.: Prentice-Hall, Inc., 1968), pp. 145–147.
3. Laurence F. Peter and Raymond Hull, *The Peter Principle* (New York: William Morrow & Company, Inc., 1969).
4. Edward Luttwak, *Coup D'Etat: A Practical Handbook* (Middlesex, England: Penguin Books, Ltd., 1969).
5. Peter F. Drucker, *Technology, Management, and Society* (New York: Harper & Row, 1970), Chapter 2.
6. *Harvard Business Review*, November–December 1969, p. 50.
7. For documentation, see Drucker's book, *The Age of Discontinuity* (New York: Harper & Row, 1969).
8. D. B. Robertson (ed.), *Voluntary Associations, A Study of Groups in Free Societies* (Richmond, Va.: John Knox Press, 1966), p. 7.
9. Robert M. MacIver, *Community, Society, and Power* (Heritage of So-

ciology Series), Leon Bramson and Morris Janowitz, eds. (Chicago: University of Chicago Press, 1970).

10. Herbert A. Simon, "The Impact of the Computer on Management," CIOS XV, Tokyo, 1969.
11. Michael J. Kami, "Systems Approach in Planning and Control," CIOS XV, Tokyo, 1969.
12. Erich Jantsch, "Technological Forecasting in Perspective" (Paris: OECD, 1967), pp. 257–270.
13. Melvin Anshen, "The Management of Ideas," *Harvard Business Review*, July–August 1969.
14. D. B. Hertz, "The Management of Innovation," *Management Review*, April 1965, pp. 49–52.
15. Arthur D. Little, Inc., *Management Factors Affecting Research and Exploratory Development*. Report for the Director of Defense Research & Engineering, Contract SD-235 (Cambridge, Mass.: ADL, 1965). Available from Clearinghouse for Federal Scientific & Technical Information under AD-618321.
16. Anshen, op. cit.
17. Ibid.

Chapter 3

1. U.S. Department of Commerce, *Technological Innovation: Its Environment and Management* (Washington: Government Printing Office, 1967); Sumner Myers and Donald G. Marquis, *Successful Industrial Innovations: A Study of Factors Underlying Innovation in Selected Firms* (May 1969): A study by the National Planning Association for the National Science Foundation. Contains an appendix of 30 selected references. MSF 69-17.
2. This discussion is based on Donald G. Marquis' brief version of the NSF study in *Innovation*, No. 7 (1969). Published by Technology Communication, Inc., St. Louis, Mo.
3. James R. Bright, *Research, Development, and Technological Innovation* (Homewood, Ill.: Richard D. Irwin, Inc., 1964).
4. Donald A. Schon, *Technology and Change* (New York: Delacorte Press, 1967), p. 65.
5. Paul R. Lawrence, "How to Deal with Resistance to Change," *Harvard Business Review*, May–June, 1954.
6. U.S. Department of Commerce, op. cit.
7. J. Jewkes, D. Sawers, and R. Stillerman, *The Sources of Invention* (New York: St. Martin's Press, 1959).
8. D. Hamberg, "Invention in the Industrial Laboratory," *Journal of Political Economy*, April 1963, pp. 96–98.

9. M. J. Peck, "Inventions in the Post-War American Aluminum Industry," *The Rate and Direction of Inventive Activity: Economic and Social Factors* (Princeton, N.J.: National Bureau of Economic Research, 1962), pp. 279–292.

10. D. Hamberg, op. cit.

11. J. L. Enos, "Invention and Innovation in the Petroleum Refining Industry," *The Rate and Direction of Inventive Activity* (National Bureau of Economic Research; distributed by Princeton University Press, 1962), pp. 299–304.

12. Everett M. Rogers, *Diffusion of Innovations* (New York: The Free Press, 1962).

13. See also Arthur D. Little, Inc., *Patterns and Problems of Technology Innovation in American Industry*, Report PB 181573 to National Science Foundation, 1963.

14. Clarence Danhoff, "Observations on Entrepreneurship in Agriculture," *Change and the Entrepreneur*, Harvard Research Center in *Entrepreneurialship History* (Cambridge, Mass.: Harvard University Press, 1949).

15. Donald A. Schon, *Innovation*, No. 6 (1969).

16. Everett Rogers, quoted by Schon, op. cit., p. 44.

17. Warren Bennis, *The Temporary Society* (New York: Harper & Row, 1968). See also "I Say Hello, You Say Goodby," *Innovation*, No. 1 (May 1969).

18. S. Colum Gilfillan, "The Prediction of Change," *The Review of Economics and Statistics*, November 1952, pp. 368–385.

19. Arthur D. Little, Inc., *Management Factors Affecting Research and Exploratory Development* (Cambridge, Mass.: ADL, 1965).

20. "Innovators of Management," *Dun's Review*, August 1969, p. 41.

21. "The Management Style of R. S. Morse," *Innovation*, No. 1 (May 1969), pp. 56–65.

22. "3M," *Innovation*, No. 5 (September 1969).

23. Jack A. Morton, "The Innovation Process in the Bell System." Report prepared under the Solid State Affiliates Program at Stanford University, November 1965. See also Jack A. Morton, *Organizing for Innovation—A Systems Approach to Technology Management* (New York: McGraw-Hill, 1971).

Chapter 4

1. This discussion is taken in part from the paper "Directive Management Within a Permissive Framework," by Robert K. Mueller, given at M.I.N.D. Conference on Overseas Development Effectiveness, Session III: Planning and Collaboration, Colby Junior College, New London, N.H., September 26–28, 1969.

2. Peter F. Drucker, "Long-Range Planning," *Technology, Management, and Society* (New York: Harper & Row, 1970).
3. Alex Bavelas, "Communication Patterns in Task-oriented Groups," in *The Policy Sciences*, edited by D. Lerner and H. L. Lasswell (Palo Alto: Stanford University Press, 1951), pp. 193–202.
4. Harland Cleveland and Harold D. Lasswell (eds.), *Ethics and Bigness* (New York: Harper & Row, 1962), p. xvi.
5. Virginia A. Carollo, *The Voluntary Sector in World Development: An Emerging Profession* (New York: Management Institute for National Development, November 1958).
6. *Playboy,* March 1962. The subsequent discussion of forecasting is based on this article.
7. Erich Jantsch, *Technological Forecasting and Perspective* (Paris: OECD, 1967).
8. Arthur D. Little, Inc., *Management Factors Affecting Research and Exploratory Development.* Report for the Director of Defense Research & Engineering (Cambridge, Mass.: ADL, 1965).
9. Conference held at Hanover Technical Fair, Federal Republic of Germany, April 1966.
10. *Research and Development in American Industry* (New York: McGraw-Hill, Department of Economics, May 6, 1966). The results of a survey of U.S. industry concerning the structure of industrial R&D in 1966 and 1969 for 15 industrial sectors and for industry as a whole.
11. James Zwolenik, *Technology in Retrospect and Critical Events in Science.* Contract NSF C 535, Vol. I, December 15, 1968. Updated in Volume II (1969) by The Illinois Institute of Technology for the National Science Foundation.
12. Arthur D. Little, Inc., *Patterns and Problems of Technical Innovation in American Industry.* Report PB 181573 to National Science Foundation (Cambridge, Mass.: ADL, 1963).
13. Burton V. Dean, *Evaluating, Selecting, and Controlling R&D Projects,* AMA Research Study 89 (1968).
14. William K. Hodson, "R&D Costs: When to Blow the Whistle," *Columbia Journal of World Business,* September 1969, pp. 92–96.

Chapter 5

1. Lord Ritchie-Calder, "Mortgaging the Old Homestead," *Foreign Affairs,* January 1970.
2. Committee on Science and Astronautics, U.S. House of Representatives, July 1969.
3. Barnaby C. Keeney, "The Humanities: Episode or Continuum," *Educational Record,* Summer 1969, pp. 256–260.

4. G. Burck, "Knowledge: The Biggest Growth Industry of Them All," *Fortune*, November 1964, pp. 128–131.
5. B. Lamar Johnson, *Islands of Innovation Expanding: Changes in the Community College* (Beverly Hills, Calif.: Glencoe Press, 1969), p. 5.
6. "Technological Trends in Major American Industries," U.S. Department of Labor Bulletin 1474, February 1966.
7. *SIPRI Yearbook of Armaments and Disarmaments, 1968/69* (New York: Humanities Press, Inc., 1969).
8. U.S. Department of Labor, op. cit.
9. Herman Kahn and Anthony J. Wiener, *The Year 2000* (New York: The Macmillan Company, 1967).
10. Summarized in Lord Ritchie-Calder, op. cit.
11. U.S. Department of Commerce, *Technological Innovation: Its Environment and Management* (Washington: Government Printing Office, 1967).
12. "New Rules for the Seventies," European Research Report No. 69-2, November 1969.
13. R. K. Mueller, Chapter 4, *Risk, Survival, and Power* (AMA, 1970).
14. Arthur D. Little, Inc., *Program Evaluation of the Office of State Technical Services.* Report to Department of Commerce, October 1969.
15. Louis R. Pondy, "Effects of Size, Complexity, and Ownership on Administrative Intensity," *Administrative Science Quarterly*, March 1969, pp. 47–61.
16. Pondy, op. cit.

Chapter 6

1. Theodore Levitt, "Marketing R&D for Marketing Innovation," *Chemical and Engineering News*, October 16, 1961.
2. Will Durant and Ariel Durant, *The Lessons of History* (New York: Simon & Schuster, 1968), p. 36.
3. Thomas Edison, "The Dangers of Electric Lighting," *North American Review*, November 1899.
4. Rolf P. Lynton, "Linking an Innovative Subsystem into the System," *Administrative Science Quarterly*, September 1969, pp. 398–415.
5. W. W. Cooper, H. J. Leavitt, and M. W. Shelly, III, *New Perspectives in Organization Research* (New York: John Wiley & Sons, 1964). See especially Chapter 4 by H. J. Leavitt for a fuller discussion of complex organizations.
6. Edward Roberts, "What It Takes to Be an Entrepreneur . . . And to Hang onto One," *Innovation*, No. 7 (1969), p. 47.
7. Victor A. Thompson, "How Scientific Management Thwarts Innovation," *Trans-Action*, June 1968, pp. 51–56.

8. William W. McKelvey, "Expectational Non-Complementarity and Style of Interaction Between Professional and Organization," *Administrative Science Quarterly*, March 1969.
9. Derived from Talcott Parson, *The Social System* (New York: The Free Press, 1951).
10. Mack Hanan, "Corporate Growth Through Venture Management," *Harvard Business Review*, January–February 1969.
11. Parson, op. cit.; Lynton, op. cit., pp. 398–415.
12. Hanan, op. cit.

Chapter 7

1. "The Chief Executive — And His Job," *Studies in Personnel Policy No. 214* (New York: The Conference Board, 1969).
2. Michael J. Kami, "Systems Approach in Planning and Control," CIOS XV, Tokyo, 1969.

Chapter 8

1. Graduate School of Business, University of Chicago, 1962.
2. Willard F. Mueller, "Origin of Du Pont's Major Innovations, 1920–1950," in James R. Bright (ed.), *Research, Development, and Technological Innovation* (Homewood, Ill.: Richard D. Irwin, Inc., 1964), p. 383.
3. *Características de los Institutos Latino-Americanos de Investigación Technológica*, Organization of American States, Department of Scientific Affairs, 1965.
4. Jorge A. Sabato, "Quantity Versus Quality in Scientific Research: I. The Special Case of Developing Countries," *Impact of Science on Society*, Vols. XX and XXIII (Paris: UNESCO, 1970).

Chapter 9

1. For a good discussion of this, see Robert Kolodney and Gabriel Pepino, "Venture Capital for Entrepreneurs," *European Business*, October 1968, pp. 18–24.
2. Stanley M. Rubel, "How the Small Investor Can Intelligently Participant in Venture Capital Investing," *Investment Dealers' Digest*, June 16, 1970.
3. Harry Levinson, "On Being a Middle-Aged Manager," *Harvard Business Review*, July–August 1969, pp. 51–90.

4. Clarence Zener, "Statistical Theories of Success," *Proceedings of the National Academy of Sciences,* May 1970.

5. *Boston College Management Seminar on New Business: Innovative Technology, Management and Capital,* May 22–23, 1969, pp. 121–122 (Chestnut Hill, Mass.: School of Management, Boston College Press, 1969).

6. Kolodney and Pepino, op. cit.

7. U.S. Department of Commerce, *Technological Innovation: Its Environment and Management* (Washington: Government Printing Office, 1967), p. 21.

8. Ibid.

9. For an interesting discussion in lay terms of the United States' approach to investing capital in technologically based companies, see Arthur C. Merrill, *Investing in the Scientific Revolution* (Garden City, N.Y.: Doubleday & Company, Inc., 1962). Mr. Merrill is an investment advisory department officer with The First National City Bank of New York.

10. U.S. Department of Commerce, op. cit., pp. 44–45.

11. See *The Innovators,* written by members of the *Wall Street Journal* staff (New York: Dow-Jones Books, 1968).

12. See *Reports of the International Conferences of Ministers of Science of OECD Countries,* Paris, 1963, 1965, 1966; see also *Applied Science and Technology,* a report to the Committee on Science and Astronautics, U.S. House of Representatives, by the National Academy of Sciences, June 1967.

13. *Gaps in Technology: Electronic Components,* OECD, Paris, March 11 and 12, 1968.

14. *Technological Innovation in Britain,* report of the Central Advisory Council for Science and Technology (London: Her Majesty's Stationery Office, July 1968).

15. J. Ben-David, *Fundamental Research in the Universities* (Paris: OECD Report, 1968).

16. Private translation of "The Owners of Capital: Where Are They?" by Abd Allah 'Abd ar-Rahman al-Jifri, *UKAZ,* No. 1794, 7 Sha'ban 1390, October 8, 1970.

17. "Post Offering Experience of New Securities Issue," *Journal of Business,* January–February 1970.

18. William Rotch, "The Pattern of Success in Venture Capital Financing," *Financial Analysts Journal,* September–October 1968, pp. 141–147.

Index

Adams, Charles Francis, 62
Adams, Robert M., 64, 65
ADELA, 186–187
adoption, defined, 54
Aero Engineering Review, 85, 86
Aeronautical Science Institute, 85, 86
Agency for International Development (AID), 203
Alcoa Corp., 124
Allahabad Bank, 204
Allstate Insurance Corp., 177
American Institute of Aeronautics and Astronautics, 86
American Language (Mencken), 116
American Management Assoc., 91, 92
American Research and Development Corp., 175, 178, 180, 181, 189, 196
American Rocket Society, 86
American Telephone and Telegraph Co., 137
Anshen, Melvin, 39, 40, 41
ANVAR, 189
Aramco, 197
Archimedes, 17
Aristotle, 17, 115
Armstrong, Edwin, 51
Ashby, Sir Eric, 43
Associated Trust Holdings (ATH), 190

Astronautics and Aeronautics, 86
Atlantic Community Development Group for Latin America (ADELA), 186–187
Aurelius, Marcus, 7, 153
Australian Innovation Corp. (AIC), 203

Bacon, Francis, 115
Banco Industrial, 186
Bank Development Loan Fund, 203
Bank of America, 176, 202
Bank of Baroda, 204
Bank of India, 204
Bank of Maharashtra, 204, 205
Bavelas, Alex, 77
Bell-Rivlin Panel on Social Indicators, 10
Bell System, 187
   innovativeness at, 69–70
Bell Telephone Laboratories, 57, 69, 70, 71
Bennis, Warren, 55
Bernard, Claude, 114
Bess, William T., Jr., 60
Bessemer Securities, 180
Bolt, Beranak and Newman, 137
boredom, innovation and, 5
Brandenberger, Jacques, 51
Bright, James R., 46
Brooke, Roger, 190
Brown University, 99